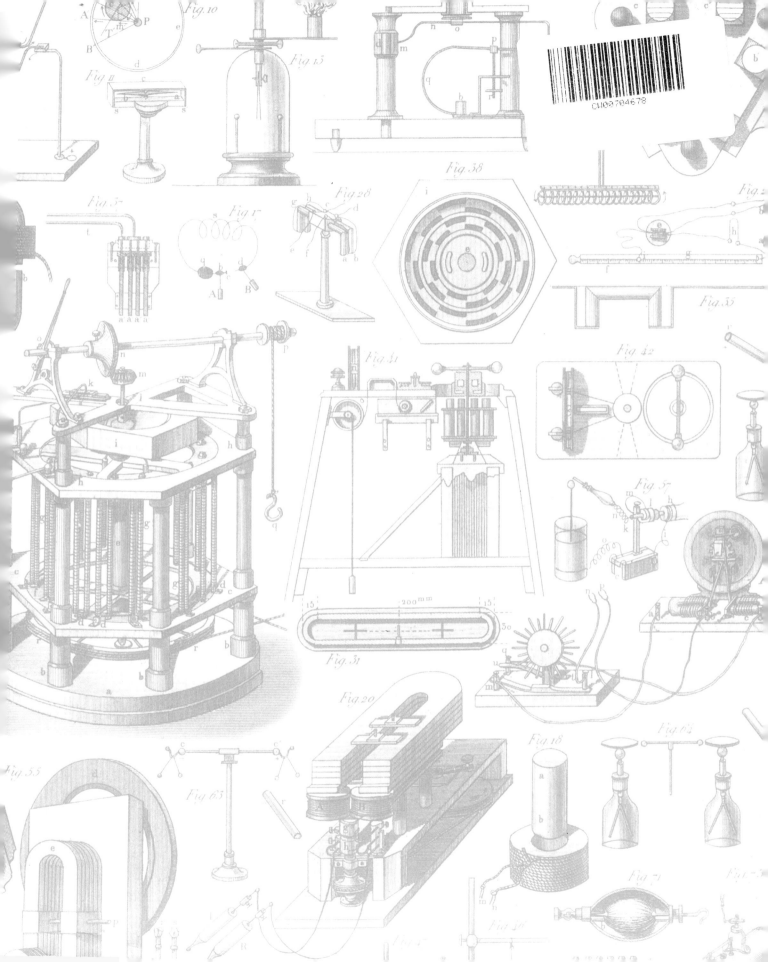

WHERE DISCOVERY SPARKS IMAGINATION

A Pictorial History Presented by
The American Museum of Radio and Electricity

ISBN-13: 978-0-9794569-0-9

Library of Congress Control Number: 2009901784
Second Printing

Editing, design and layout	John Jenkins
Copy	Diana Sheiness
	John Jenkins
	Sam Spencer
	Susan Tive
	Tana Granack
Photography	John Jenkins
	Sam Spencer
Object labels	John Jenkins
	Tana Granack

ISBN: 978-0-9794569-0-9

Published by The American Museum of Radio and Electricity, Bellingham, Washington.

American Museum of Radio and Electricity
1312 Bay Street
Bellingham, WA 98225
Phone: 360.738.3886
Fax: 360.733.2532

www.amre.us

Printed and bound in Hong Kong

ACKNOWLEDGEMENTS

The publication of this book would not have been possible but for the efforts of several individuals:

Diana Sheiness and **Sam Spencer**, for their tireless research, writing and editing of the historical narrative. Sam deserves special thanks for his patience, as what was initially a simple brochure turned into a full-blown book project. Sam also supplied several beautiful photographs, including the subtle photos of the museum facade and vacuum tubes on pages 2 and 188, respectively.

Thanks to **Susan Tive** for her talented work as wordsmith of the gallery introductions, and to **Tana Granack** for his interesting, creative and accurate object descriptions.

Jonathan Winter, **Frank Ordway**, and **Carl Nemeth** deserve a special thank you for their encouragement and support.

Rare Geissler Tube
19th century

An exceptionally rare multi-bulb Geissler
tube containing four different fluorescent
fluids: Rhodanine, Eosine, Flourecine, and
Authacein. German, c. 1890.

This tube is but one example of the many
beautiful Crookes and Geissler tubes on
display at the Museum.

CONTENTS

Oscar Wuertz instructing a youngster in the art of making his own wireless set in a Brooklyn radio shop, about 1922. (Courtesy Library of Congress)

PREFACE

Someone once said there is a fine line between collecting and insanity. If that is truly the case, then I began to lose my mind at the age of thirteen, the year I discovered an old radio in my grandparents' basement. With their permission, I took the set home and got it working. From that moment I was hooked on antique radios and other interesting objects related to the history of electricity.

Somehow I managed to discover the book "Vintage Radio" by Morgan McMahon, the only book on old radios at the time. I would spend hours poring through its pages, dreaming that someday I might own some of these amazing devices, the technological breakthroughs of their time.

During the next 40 years, I brought together thousands of artifacts and books beginning in the 16th century with the earliest investigations into electricity, and following a thread of discovery including the electrochemical battery, electromagnetism, the telegraph and telephone, electric light, and ultimately leading to the development of wireless telephony, more commonly known as radio.

In the 1990s I created a Web site as a way to share my collection with others. Like most collectors I dreamed of having a museum, and this virtual museum seemed like a good alternative. Over the years the site has grown substantially and today it is probably the largest collection of electrical and radio apparatus on the internet.

There was one problem, however: it wasn't especially portable. I found that despite the advantages of a web site, I still needed a hard copy that I could carry around and show to people. So I sat down with Microsoft Publisher and over a period of weeks created my "Sparkmuseum book" - essentially a book version of my web site. It was great to have, and I referred to it constantly. I also noticed something else: everyone who saw it wanted a copy.

Which brings me to today, to this book. When I decided to undertake this project I knew I didn't want to just print a book full of pictures. I also didn't want to do just another history book about radio and electricity. Really what I wanted was to recreate for people that sense of magic and discovery that I felt as a child when I sat down with McMahon's Vintage Radio: a sense of appreciation for the amazing minds and hands of the scientists, inventors and craftsmen whose dedication, persistence and plain hard work have made our modern world possible.

I hope you will enjoy it.

John Jenkins
Woodinville, WA
January, 2009

ABOUT THE MUSEUM

The amazing objects you see featured in this book are only a small sample of the larger, more complete collection residing at the American Museum of Radio and Electricity. Lightning strikes several times a day at the Museum, where thousands of students and visitors experience the marvelous history, science and power of electricity.

The Museum is a center for education and enlightenment—a place where students can get charged about science and discovery while surrounded by one of the most significant and complete collections of its kind in the world.

Sam Spencer Photo

THE GALLERIES

The Museum offers an exciting and educational journey through six interactive galleries spanning four centuries of scientific achievement and discovery. Each gallery focuses on a particular phase of scientific investigation and everywhere you look you'll see rare and compelling artifacts.

The Dawn of the Electrical age (1600 – 1800)	*What is electricity?*
Electricity Sparks Invention (1800 – 1879)	*What can we do with electricity?*
The Wireless Age (1880 – 1920)	*Development of wireless telegraph and telephone*
Radio Enters the Home (1920 – 1927)	*The beginnings of broadcast radio*
The Golden Age of Radio (1928 – 1950)	*Radio becomes home entertainment*
The Jones Gallery (1903 – 1980)	*The evolution of vacuum tube technology*

Following a continuous thread of invention and discovery, the Museum collection contains a wealth of unique and rare artifacts dating from the earliest days of scientific electrical experiments in the 1600s through the 1940s and the Golden Age of Radio. Artifacts from the laboratories of the early pioneers of electricity, from magnets and Leyden jars to Edison light bulbs, magnificent vacuum tubes and telephones, all are well represented.

Over 1,000 radios are on display, ranging from the early "Hertzian-wave" devices, to a complete set of early Atwater Kent "breadboards," all the way to scores of exceptional and beautifully crafted floor and table-top radios. The collection also includes rare music boxes, early phonographs, and many examples of radio broadcasting technology and memorabilia from the best-known radio companies and broadcasters.

Other rare pieces include the largest collection of 19th century electromagnetic apparatus found in any private collection, and rare and original books, treatises and scientific papers by such authors as Gilbert, Newton, Galileo, Franklin, Volta, Hertz, and Marconi. These texts illustrate the crucial steps and turning points in the development of electricity and radio.

The Museum offers visitors a first-hand introduction to the wonderful world of electricity and radio, providing every opportunity to discover, test and be amazed. In addition to the six galleries, the Museum features other popular interactive displays, including various Tesla Coils, which create dazzling bursts of lightning on command.

Visiting the Museum can make your hair stand on end! | A young virtuoso tries her hand at playing the Theremin

Adventurous visitors are invited to experience hair-raising activities in the Museum's Static Electricity Learning Center, or even create other-worldly sounds from the Theremin (the first electronic musical instrument), and much more.

Scholars and serious collectors can examine one of the most complete collections of early original recordings of popular music ever amassed. Over 30,000 original recordings, many of which have been painstakingly digitized, cataloged and documented. Highlights from these educational and historic recordings are broadcast regularly on KMRE 102.3 FM, the Museum's own radio station, and streamed on the Internet at: www.amre.us

KMRE-LP 102.3 FM

KMRE 102.3 FM, "The voice of the American Museum of Radio and Electricity," is an independent, community-based Low Power radio station owned and operated by the Museum since 2004. KMRE is a powerful and entertaining extension of the Museum, broadcasting some of the finest educational, cultural and historically significant recordings ever produced, and streamed on the Internet to listeners all over the world.

In addition, the Museum print library consists of tens of thousands of science and history books, early radio themed magazines, technical manuals, and sheet music ranging up to the 1950s. The Museum even has its own radio repair shop specializing in analog radios and equipment.

Curator Jonathan Winter adjusts the high-frequency electrostatic display on a rare combination Oudin and Tesla coil. The device was produced around 1910 by the Macintosh corporation of San Francisco and was used to power early electro-medical apparatus, including X-ray tubes.

Visit the Museum online at: WWW.AMRE.US

SP⚛RK Science-Powered Adventures
for Real Kids
a program of the AMERICAN MUSEUM of RADIO and ELECTRICITY

The spark of discovery is alive and well at the American Museum of Radio and Electricity. The Museum offers a variety of educational activities guaranteed to ignite the curiosity of students and visitors of all ages.

Using its unmatched collection of electrical inventions and related artifacts as inspiration, the Museum has created a dynamic education program that sparks the imaginations of over five thousand boys and girls annually. In partnership with Western Washington University and local school districts the Museum has created a program that both supports the curriculum in schools and expands opportunities for students to get interactive experiences with science. SPARK ignites young imaginations through hands-on Science Saturday classes, electronics workshops, school assemblies, special focus tours even science summer camps, primarily focusing on the process of innovation, invention and the core facets of radio, electricity and physics. Presentations are performed at the Museum and on-site in the schools.

SPARK offers educators priceless resources to augment their regular science curriculum. The Museum offers special courses to increase grade school teachers' comfort and proficiency in teaching science. Teachers taking part in this program receive continued education credits to fulfill key professional development goals. The Museum also offers internships for science and history students through-out the area.

Curator Jonathan Winter entertains with the Tesla Coil

Dad and the kids enjoy Science Saturday

SUPPORTING THE MUSEUM

The American Museum of Radio and Electricity is a place where students, families and friends can experience the profound history and science of electricity. Gifts and contributions ensure the Museum continues to be an exciting and dynamic place – where every visit is a new experience to enjoy and learn from. Your tax-deductible donation supports and sustains the Museum's science education program, interactive galleries, exhibit spaces and displays. Your generosity allows the Museum to continue electrifying thousands of visitors every year with inspiration and a sense of discovery everyone should know. Please contact 360.738.3886 or donations@amre.us

ELECTRICAL ORRERY. Representations of the sun, earth and moon in painted wood and bone. All three are balanced on two brass wires and rotate when exposed to static electricity from a static generator. This electrostatic philosophical accessory is pictured in Benjamin Pike's 1848 Catalog Vol I. fig 316, p. 282; also pictured and described in numerous Natural Philosophy books and other trade catalogs of the period; pictured and described in the King George III Collection as well.
American, c. 1855

GALLERY ONE
The Dawn of the Electrical Age

The Dawn of the Electrical Age

Ever wonder why your socks stick together when they come out of the dryer?

In BC 600, the ancient Greeks were also wondering about the mystery of electricity and what causes those bright flashes of light in the fur cloth they used to polish jewelry.

For the next 2000 years, electricity is believed to be a "magic power" originating from a divine source. Then, during The Age of Enlightenment in 18th century Europe, the rebirth of logic and reason inspire curiosity and new questions about the natural world.

Innovative thinkers like Benjamin Franklin eagerly explore electricity, pioneering original and often dramatic experiments to find out more.

Old explanations and fears are slowly overcome as thinkers from around the globe uncover exciting information about the nature of electricity.

A world of possibilities is born!

GALLERY ONE
THE DAWN OF THE ELECTRICAL AGE

MAGNETISM AND STATIC ELECTRICITY were known to exist long before their relationship became appreciated and exploited. However, with the exception of the magnetic compass, these phenomena made few contributions to the everyday lives of people until centuries after they were first observed.

MAGNETISM

Civilized man has long been aware of magnetism. The earliest record goes back to 900 BC, when a Greek shepherd reportedly discovered that the nails in his shoes were attracted to the stones in a certain field. The shepherd told others of this strange observation, and today he is still remembered in the term "magnetic," a word derived from his name, "Magnus." The stones in Magnus' field were naturally magnetized ferrites, or "lodestones." Lodestones were used by the Chinese as direction pointers by at least the 8th Century, and perhaps far earlier. By the 1400s, lodestones were in use in navigational devices on ships in many parts of the world. Magnets were put to few practical uses besides compasses until much later when it was finally discovered that a magnetic field could induce an electric current in a wire.

No one person is entitled to all the credit for uncovering the secrets of electricity and its relationship to magnetism. Rather, progress in this field occurred in small increments until the pieces finally fell into place.

Lodestone - 18th Century (English)

ELECTRICITY

Electricity was first recognized in the form of electrostatic charges. An electrostatic charge consists of a buildup of electrons on the surface of an insulated body. Such a charge can be created by rubbing one body against another, such as the soles of one's shoes across the surface of a carpet. When this happens, one of the bodies becomes negatively charged and the other becomes positively charged. The electrostatic charge is released if a charged body comes in contact with another object that is oppositely charged or grounded. Lightning, for example, is a released electrostatic charge.

The person credited with discovering electrostatic charges was Thales of Miletos (624-546 BC). Thales observed that when amber was rubbed against cat fur, the amber attracted bits of feather.

William Gilbert   (1544 - 1603)

Further progress in the understanding of magnetism and electricity came from an English physician, William Gilbert (1540-1603), who discovered in 1600 that the earth itself is a magnetic body and that this was why compasses point north. Gilbert also is attributed with coining the term "electricity," which he derived from "elektron," the Greek word for amber. The unit of magnetomotive force, the "gilbert," was named after him.

In 1729, an English investigator, Stephen Gray (1666-1736), discovered that electrostatic charges reside on the surfaces of the objects used to produce the charge, and also made the far more significant discovery that electrostatic charges can be conveyed to a different location. Gray tested numerous materials, including various household items, metals and threads, and found that most of them could serve as conduits for electricity. He determined whether or not a transfer of electricity was successful by testing whether the receiving object acquired the ability to pick up bits of feather or brass leaf.

Gray's experiments comprised a rudimentary demonstration of electrical conductance. In fact, Gray believed that electricity is a fluid, which is not so different from today's understanding of this form of energy.

Charles Francois du Fay (1698 – 1739) discovered in 1733 that electricity comes in two flavors, which he called "resinous" and "vitreous." These terms were later redefined by the American patriarch, Benjamin Franklin (1706 - 1790) as "negative" and "positive," the terms by which we know them today. Franklin was also the first to report that positive and negative charges are attracted to one another, and that like charges are mutually repelled.

Benjamin Franklin also is attributed with showing that lightning was electrical in nature. He is well-known for his famous experiment in which he flew a kite with a key attached to its string during a thunderstorm. Contrary to popular belief, the kite was not struck by lightning, but Franklin did observe the buildup of electrostatic charges on the metal key. This enabled him to postulate that the charges that built up in clouds were electrostatic in nature, and that lightning itself was a type of electrical discharge.

The understanding Franklin gained from his experiments with

electricity enabled him to invent the lightning rod, which is a grounded iron rod that is attached to a tall building, extending above the roof. The rod safely discharges the static electricity from the charged air directly into the ground without harming the building or its occupants. Before lightning rods came into use, it was a common practice for bell ringers to be sent up into church towers to ring the bells during a storm in order to frighten away the lightning. As a result of this misguided practice, bell ringers were common victims of lightning strikes. No doubt the bell ringers were quite pleased when lightning rods came into use.

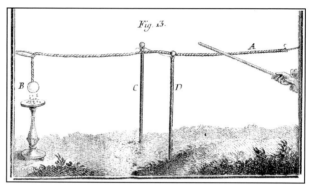

Electrical conduction experiment of Stephen Gray
Source: Saggio intorno all'Elettricita de Corpi, Nollet

Magnetic Toy

The toy consists of a mahogany box containing four numbered ebony chips. The chips can be removed and replaced in any order. Each chip contains a hidden iron bar, situated parallel to the face of the chip and each one oriented differently from the others.

The toy is missing its "key", shown in the illustration on the right. The key contains four paper disks which are mounted like compass needles and are free to rotate. To operate the puzzle, the host averts his eyes and invites a guest to rearrange the ebony chips in any order and close the case. The host then places the "key" on top of the still closed puzzle and reads the correct answer from the key, amazing his guests. What the guest can't see is a small iron bar attached to each disk inside the key that causes the disk to align with the numbered ebony chip in the puzzle. The magnetic attraction causes the iron bar to align with the chip, bringing the appropriate number to view in the window.

The mahogany box measures 10-3/4" x 2-3/4" Unsigned, but very similar to item number E71 described and illustrated on p. 445. of Public and Private Science, The King George III Collection, Morton and Weiss, London 1993.

Magnetic Toy
c. 1842 or 1824

Magnetic Toy key- George Adams
From the King George III Collection
From "Public & Private Science" by Morton & Wess, London

ELECTRO-STATIC PLANETARIUM. Unsigned,. Italian, c. mid-19th century

The first devices for producing electricity were very simple. The ancient Greeks discovered the strange effects of amber rubbed with fur and other material. In the 17th century, scientists used sticks of resin or sealing wax, glass tubes and other objects. By the time of Benjamin Franklin, large glass tubes about three feet long and from an inch to an inch and a half in diameter were popular. These were rubbed either with a dry hand or with brown paper dried at the fire.

There are two major categories of electrical machines: Friction and Influence. A friction machine generates static electricity by direct physical contact; the glass sphere, cylinder or plate is rubbed by a pad as it passes by. Influence machines, on the other hand, have no physical contact. The charge is produced by inductance, usually between two or more glass plates spinning in opposite directions.

All through the 18th and 19th centuries there was tremendous interest in electricity. Scientists such as Franklin, Nollet, Coulomb, Volta, Oersted, Ampère, Ohm, Faraday, Joule and others made major advances. Prior to the invention of the induction coil in 1831 however, the only way to generate high voltage electricity was via static generators such as these.

In devices similar to the Winter Friction Machine shown opposite, rotating the wheel created a static charge, which was available on the "prime collector" (the brass ball or cylinder at the top or front of the device.) The charge could then be stored in a Leyden jar or measured with an electroscope.

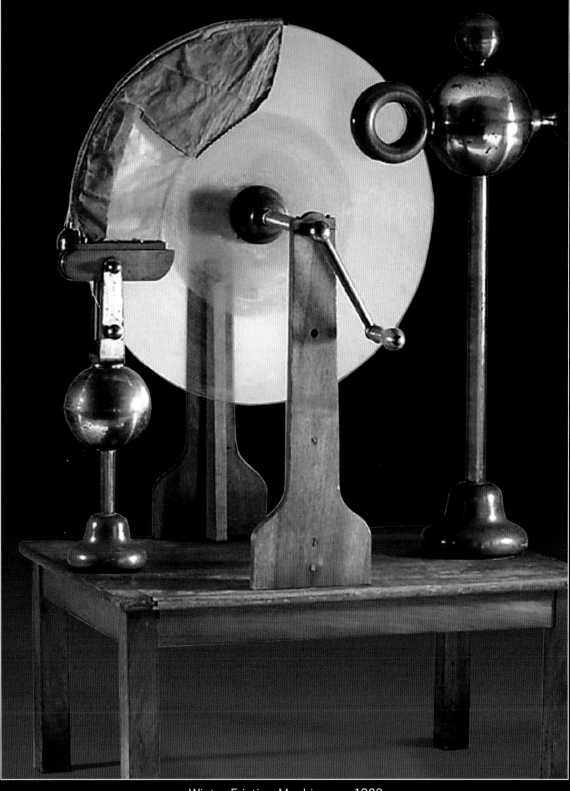

Winter Friction Machine   c. 1900

RAMSDEN FRICTION MACHINE. Unsigned. In 1768 Jesse Ramsden (1735—1800) constructed his classic glass-plate electrical machine. The museum has two 19th century Ramsden machines on display. The one pictured here was restored by Museum curator John Jenkins. American, c. 19th Century.

PLATE ELECTRICAL MACHINE. Signed "J.M. Wightman." Walnut cruciform base design with original glass plate and large brass conductors mounted on thick and hollow hand blown glass insulators; Pictured in Wightman's Catalog (1842). Came with assortment of other philosophical instruments all made and some signed by Wightman. Very fine condition. American, c. 1842.

NAIRNE FORM FRICTION MACHINE. Unsigned. Edward Nairne (1726—1806) was one of the first to build an electrical machine using a glass cylinder instead of a sphere. The machine at right, on display at the Museum, is a smaller version of Nairne's own machine. The glass cylinder was 19 inches long and 12 inches in diameter, and the main conductor (the metal cylinder) was one foot in diameter and five feet long! American, c. 1850.

RAMSDEN FRICTION MACHINE. Unsigned.
This is the second Ramsden machine on display at the
museum. Original condition. French, c. 19th Century

TÖPLER-VOSS INFLUENCE MACHINE. Unsigned, but
likely "A. Dall'Eco, Firenze Viale Principe Eugenio."
Walnut ovoidal base supported by wooden feet measures
$15^{1/2}$ x 11". Larger glass disk measures $12^{1/2}$" dia, smaller
disk is $10^{1/2}$" dia. Overall dimensions are 15" tall x 21" in
length, noting variation in length due to belt-tensioning
mechanism.

A predecessor to the modern Whimshurst machine, this
machine was derived from the mechanical improvements
made between 1865 and 1880 by the Russian physicist
August Töpler, the German physicist Wilhelm Holtz
(1836-1913) and J. Robert Voss, a mechanician from
Berlin. The machine is very efficient and produces a
large spark. Italian, c. 1882.

INFLUENCE MACHINE, Unsigned, but likely Edward S.
Ritchie Co., Boston, MA. Features two counter-rotat-
ing disks and four large Leyden jars. Measures 21 x
27" at mahogany base, 28" High. American, c. 1880.

Leyden Jars
The "Electric Phial"

The Leyden jar originated about 1746 through the work of Dutch physicist Pieter van Musschenbroek of the University of Leyden and Ewald Georg von Kleist of Pomerania, working independently.

A Leyden jar consists of a glass jar with an outer and inner metal coating covering the bottom and sides nearly to the neck. A brass rod terminating in an external knob passes through a wooden stopper and is connected to the inner coating by a loose chain.

When an electrical charge is applied to the external knob, positive and negative charges accumulate from the two metal coatings respectively, but they are unable to discharge due to the glass between them. The result is that the charges will hold each other in equilibrium until a discharge path is provided. Leyden jars were first used to store electricity in experiments, and later as condensers in early wireless equipment.

Early gold leaf Leyden Jars
1st half 19th century

LEYDEN JAR BATTERY signed Ducretet

Using a term borrowed from military artillery, this early 'battery' consists of four Leyden jars connected together to increase their capacity to store a static charge. Each jar is 12" in height and 4" in diameter, supported by a 4" diameter brass cup. The jars are connected in parallel top and bottom with brass chains and rod attachments. Each is mounted on individual 9" glass rods with brass fixtures; the entire apparatus is supported on a mahogany base.

The early natural philosophers first thought electricity was a mysterious fluid; they called the devices developed to capture that fluid, condensers. Later, when it was discovered that electricity was a stream of electrons the condenser became known as the capacitor.

French, c. 1865

FRANKLIN'S BELLS

During one of his many experiments in the mid-18th century, Benjamin Franklin noticed that a pith ball or cork would initially be attracted to a charged object, but if the two touched the ball would be repelled. He realized that this was because the two objects were initially charged differently (dissimilar charges attract) but once they touched they took on the same charge (similar charges repel.)

Franklin realized that if he replaced the charged object with a bell, he could make an "electric bell." He soon found practical use for his bell as a lightning detector. When connected to his lightning rod, the bell would ring whenever an electrical storm was nearby.

Apparently he forgot to disconnect the bells during one of his many trips to France and his wife was quite unhappy with him! Italian, 19th Century.

ELECTRICAL SPORTSMAN. By Joseph Wightman Originally three small paper birds were attached by threads to the electrode of the jar. When the jar was charged, the static electricity made the birds seem to fly into the air.

When the sportsmen's rifle touched the jar's electrode, the jar discharged — the resulting spark sounding something like a small rifle shot — and the birds fell, hanging from their threads at the side of the jar. American, c. 1840s.

ELECTRICAL SPORTSMEN. Unsigned.
These very rare Sportsmen were used as a novelty to discharge Leyden Jars. The base is held in the hand and the barrel of the rifle is moved near the electrode of a charged jar. Eventually the jar discharges into the Sportsman with a loud "crack!" as a flash appears at the barrel of the rifle. See page 283, fig 318 of Pike's Catalog Vol. I. French, c. last qtr, 18th Century.

In the 1770s, Alessandro Volta became interested in the characteristics of swamp gases. In order to test flammability he invented his cannon, also called a spark eudiometer. He filled the cannon with methane gas or a mixture of hydrogen and oxygen, placing a cork in the top to keep the gas from escaping. A metal rod tipped with a brass ball was inserted at the bottom to serve as a spark plug.

If the gas were flammable, and the proportions right, it would explode when sparked from a Leyden jar. Volta's early cannons reportedly were capable of sending a lead ball twenty feet, denting a board.

While they were sometimes fashioned as pistols, these devices were never intended as weapons, and were hardly as practical or effective as the gunpowder weapons which were common at the time. They did, however, make for impressive demonstrations, and led to the invention

Volta Multi-Canister
c. 19th century

Six Volta canisters arranged on a spinning disk. When attached to a static machine, the canisters fire in sequence as each electrode receives an electrical charge from the machine.

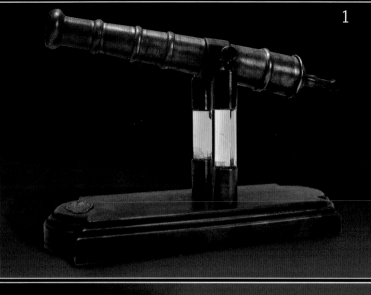

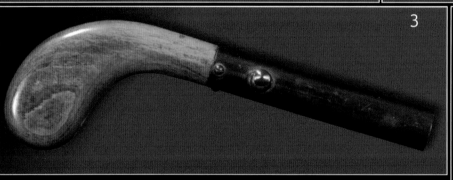

Various Volta Pistols, Cannons, and Canisters

1: Volta Cannon, c. 1860 by Angelo Arrighini. Italy

2: The elegant flask shape is 5" at its highest point. Provenance, design, and detailing of brass suggest that this objects dates from the first quarter of the Nineteenth Century and was very likely contemporary with Alessandro Volta. Italian, Very fine condition. c. 1820s

3: Volta Pistol, c. 19th century. Unknown maker & origin

4, from right to left:

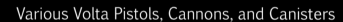

 a) Volta Canister with Bullseye made of polychromed and variously painted tin, cork glass and brass; measures 7-1/2" high and bullseye circle is 3-1/2" diameter;

 b) Volta's Canister made of above materials, measures 4-1/4" in height;

 c) Philosophical Instrument made of same painted tin and multi-colored striped barrel which fits into the main tin container. Unknown what this device was used to demonstrate but likely hydrostatics; measures 7-1/2" in height.

5. Electrical sportsman shown as used with Bullseye Canister.

Early Volta Cannon. Inscription on base reads: "Pistola du Volta 1787"
Italian, c. 1787

Scintillating Tubes

Scintillating tubes were a popular form of electric entertainment throughout the 18th and even 19th centuries. A line of small foil diamonds or discs spiraling down a hand-blown glass tube, with only a small space separating the pieces.   A dramatic electric display of sparkling light is created when a static charge is applied to the brass ball finial atop the tube -   the charge spirals down the path of tin-foil diamonds, creating a bright spark at the union of each piece.

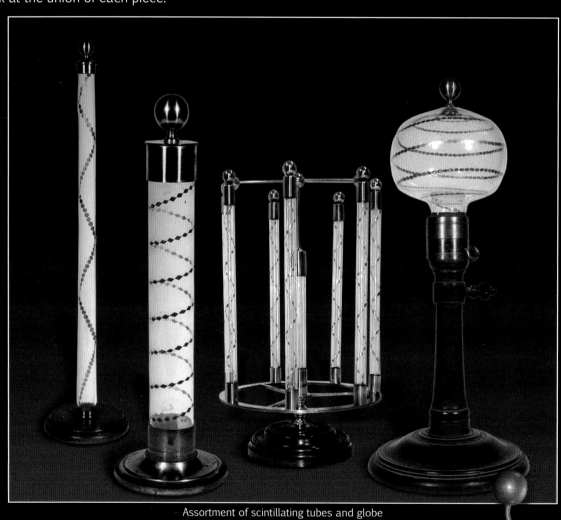

Assortment of scintillating tubes and globe
c. 19th century

Hand-held Sparkling Tube
Likely French

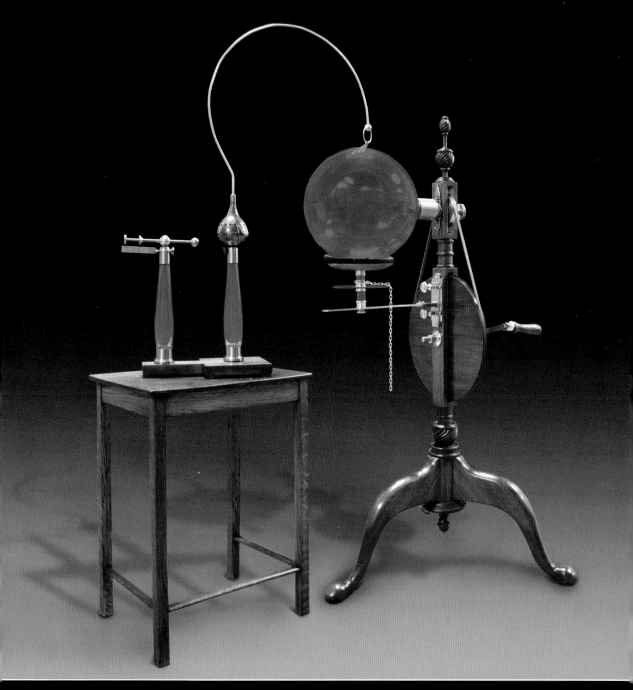

JOSEPH PRIESTLEY FRICTION MACHINE OF 1769 (REPLICA)
Best known for his discovery of oxygen, Joseph Priestley also experimented with electricity and wrote the first comprehensive history of the subject, published in 1769. The book includes detailed descriptions of several static electricity machines.

Since no complete examples of these machine survive today, Museum curator John Jenkins constructed a faithful reproduction of one machine, using the illustration and description in Priestley's book as a guide. The wood base was carved to Jenkins's specifications by Lee Grindinger, a master furniture maker from Montana. The glass was blown by Seattle artist David Smith of Blowing Sands Glass Studio and the remainder of the work was done by Jenkins. Shown here. The machine stands about three feet tall, supporting a 10" glass sphere

The Thunder House

In the late 1770s there was heated debate regarding the relative merits of a point or ball as the electrode of a lightning rod. Benjamin Franklin (the inventor of the lightning rod) supported the idea that the rod should terminate in a point. Others (primarily in England, where Franklin wasn't well liked due to his support of American independence) believed a ball would make a better spark gap. After all, most of the devices up to this time used a brass ball as an electrode, Franklin's "point" being a relatively new idea.

Dr. James Lind of Edinburgh invented the Thunder House in order to test Franklin's theory. Experiments with the device indicated that Franklin was wrong and the ball worked better, but in truth the apparatus proved nothing. On the enormous scale of nature there is little difference between the two.

How does it work?

Prior to the demonstration the operator places a small bit of explosive powder in the "exploding window."

The Leyden jar is then charged to provide the energy for the "lightning", in this case a simple spark.

Using the glass ball as a handle, the operator rotates the shorting bar until a spark jumps from the jar's electrode to the lightning rod.

If the lightning rod's ground connection is intact, the charge bleeds harmlessly away. However, if the ground is not intact, the spark ignites the powder and the window is blown away.

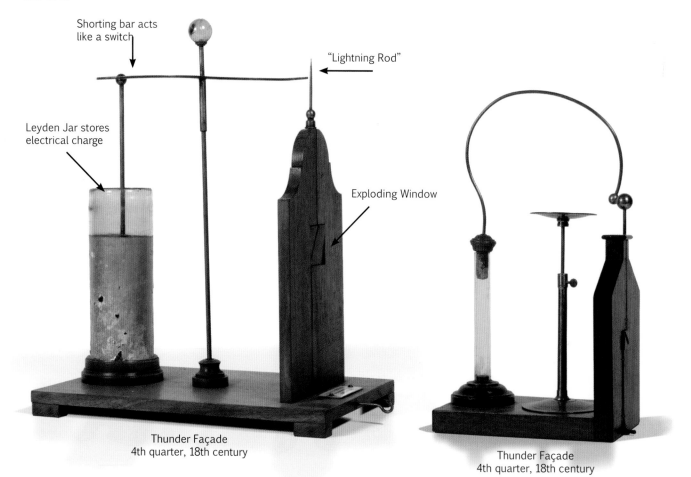

Shorting bar acts like a switch

"Lightning Rod"

Leyden Jar stores electrical charge

Exploding Window

Thunder Façade
4th quarter, 18th century

Thunder Façade
4th quarter, 18th century

Electroscopes and Electrometers

Electroscopes were the first instruments used to measure electric charge. These first "voltmeters" simply indicated the presence of charge but didn't provide a calibrated reading. Later devices, which we now call electrometers, did include a calibrated scale for reading the strength of the charge.

The first electroscope was a device called a versorium, developed in 1600 by William Gilbert (1544-1603), Physician to Queen Elizabeth I. The versorium was simply a metal needle allowed to pivot freely on a pedestal. The needle is attracted to charged bodies brought near.

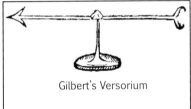

Gilbert's Versorium

A single hanging thread, called a "Pendulous thread" by Stephen Gray, (1666-1736) was introduced around 1731. The thread would be attracted to any electrified body nearby. In 1753 John Canton improved the electroscope by adding two small pith balls suspended by fine linen thread. When placed in the presence of a charged body, the two balls would become similarly charged, and since like charges repel, the balls would separate. The degree of separation was a rough indicator of the amount of charge.

In 1770, William Henley developed the first portable quadrant electrometer.

The device consisted of an insulated stem with an ivory or brass quadrant scale attached. A light rod or straw extended from the center of the arc, terminating in a pith ball which hung touching the brass base of the electrometer. When the brass was electrified the ball would move away from the base, producing an angle which could be read off of the scale.

In 1779 Tiberius Cavallo (1749-1809) designed an improved electroscope, his "pocket electrometer." The device included several improvements including, for the first time, placing the strings and corks inside a glass enclosure to reduce the effect of air currents.

The first true electrometer came from Horace Benedict de Saussure (1740-1799), who placed the strings and balls inside an inverted glass jar and added a printed scale so that the distance or angle between the balls could be measured. It was de Saussure who discovered that the distance between the balls was not linearly proportional to the amount of charge. However, the exact "inverse square" relationship between charge and distance would be left for Charles Coulomb to discover in 1784.

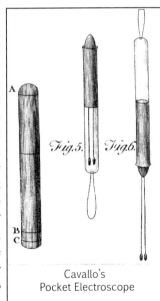

Cavallo's
Pocket Electroscope

Abraham Bennet (1750 - 1799) discovered that strips of gold foil were much more sensitive than cork or pith and created the first gold leaf electroscope in 1786.

Henley's Electrometer

Kelvin Astatic Galvanometer, c. 1890

Galvanometers were the predominant electrical measurement instruments of the 19th century. This beautiful ex- ample is an astatic mirror galvanometer first developed by Sir William Thomson (Lord Kelvin) around 1858. The device was noted for its sensitivity, being capable of detecting currents as small as 20 picoam- peres (pA).

Kelvin's galvanometers were a vital tool in the development of the transatlantic telegraph. Signals passing through these very long submarine telegraphic cables were extremely small by the time they reached the end, but Thomson's gal- vanometers were capable of detecting them.

Because this galvanometer was so sen- sitive, it was affected by the earth's mag- netic field. Thomson compensated for this by adding the curved magnet at the top of the device.

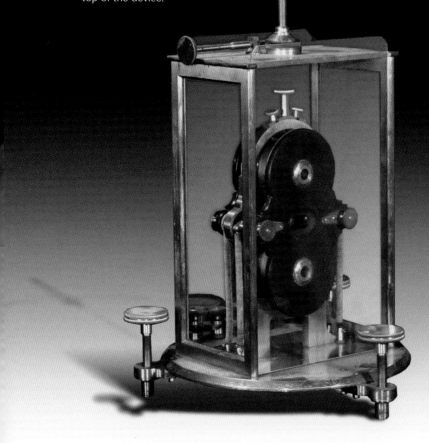

GALLERY TWO
Electricity Sparks Invention

The invention of the battery generates renewed exploration of electricity. Imagine the excitement when early inventors realize that energy is all around them, in the air and in objects. If they can collect and store it, they can put this power to work.

The discovery of electromagnetism and the invention of the electromagnet send the sparks flying fast and furious. Is it possible to use this new force to move a lever or run a mill?

Inspired by this question, scientists engage in a flurry of theories and experiments to produce the electric motor and the telegraph, paving the way for the telephone, electric lights, and even radio.

Time and distance are overcome by human ingenuity, changing forever the way people work and play.

The power of electricity is unleashed!

GALLERY TWO
ELECTRICITY SPARKS INVENTION

IN 1780, LUIGI GALVANI (1737 - 1798) found that he could cause twitches in isolated frog legs by applying static electricity to them. He also was able to elicit the twitches by hanging the dissected legs on brass rings which in turn were contacted with an iron railing. The contact between the two dissimilar metals generated a current that stimulated the frog muscles. However, Galvani mistakenly believed that his manipulations caused a current to be generated by the frog muscle itself. He named this phenomenon "animal

filled with a solution of salt. The glasses were linked together by strips of metal running from one glass to the next. This device, known as the "Crown of Cups," produced a large flow of electricity.

After this success, Volta simplified the cumbersome Crown of Cups by sandwiching silver and zinc disks with disks of wet cardboard, thus creating the first electric battery, which was called the "Voltaic Pile." This was the first

Illustration from *De Viribus Electricitatis in Motu Musculari*
by Luigi Galvani, 1791

electricity." When Galvani reported his observations, another Italian scientist, Alessandro Volta (1745 - 1827), was intrigued. However, Volta rejected Galvani's explanation for his results.

Volta believed that Galvani's electricity emanated solely from the contact of the two dissimilar metals, and not from the frog legs themselves. Testing his theory, Volta was able to produce a current in the absence of any muscle tissue. In 1800, Volta applied what he had learned to construct a device consisting of numerous wine glasses each partially

device to produce a steady stream of electricity and served as a catalyst for the explosion of electrical invention that followed. Today's name for the basic unit of electrical potential, the "volt," is derived from the name of this pioneer in electrical research.

ELECTROMAGNETISM

Within a few years after Volta performed his experiments, sufficient knowledge finally accumulated to allow the connection to be made between magnetism and electricity.

The stage for this discovery was set by the Danish physicist, Hans Christian Oersted (1777-1851). Oersted showed that if a magnetic needle were placed near a wire that carried an electric current, the needle would actually move until it was standing at right angles to the wire. This experiment demonstrated an interaction between electricity and magnetism. Though lacking any explanation for this curious observation, Oersted published his results in 1820. It would not be apparent till much later, but Oersted's discovery provided the fundamental piece needed for the eventual development of the electric motor, the telegraph, the telephone and the dynamo.

Intrigued by Oersted's observations, other scientists were drawn to investigate this startling phenomenon. Among these was the French physicist André-Marie Ampère (1775-1836). Ampere was quick to grasp the relationship between electricity and magnetism. He developed a mathematical theory of electromagnetism and formulated a mathematical law to describe the magnetic force between two electrical currents. Put simply, Ampere's law states that parallel currents attract and opposing currents repel one another. Though Ampere was not the one to realize it, his results implied that because electricity and magnetism shared the ability to attract and repel, the two forces were related. Ampere also devised an instrument, now called a "galvanometer," for measuring the magnitude of current flowing in a conductor. The name we use today for the unit of electrical current is the "amp," which is derived from "Ampere."

THE ELECTROMAGNET

In 1825, William Sturgeon (1783 - 1850) significantly

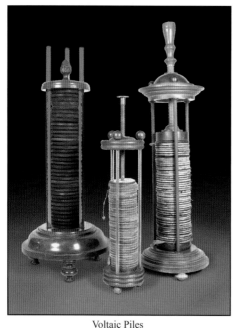

Voltaic Piles
c. 19th Century

Schweigger's Multiplier. In September of 1820, Johann Schweigger discovered that wrapping coils of wire around a compass needle dramatically increased the effect of an electric current on the deflection of the needle.

improved Oersted's moving needle experiment. Harnessing the same phenomenon that Oersted had observed, Sturgeon created the first electromagnet by wrapping a bare wire (insulated wire hadn't yet been developed) around an iron bar covered with silk, and passing a current through the wire. Sturgeon's electromagnet was the first showing that electricity could be converted into another form of energy, that is, to electromagnetic energy.

The American scientist, Joseph Henry (1797-1878), dramatically improved Sturgeon's prototype electromagnet. Henry realized that he could increase the strength of the magnetic field in an electromagnet by increasing the length of the wire. To accomplish this, he wound many coils of wire around the iron bar, insulating each coil from the next with silk thread. This ensured that the current would pass linearly through the entire length of the wire. When the wire was connected to a Voltaic pile, the resulting electromagnetic field was far stronger than the field created by Sturgeon's device.

Henry's Electromagnet
(Photo courtesy of the Smithsonian Institution)

THE ELECTRIC MOTOR

One of the major developments that emerged from the marriage of magnetism and electricity was the electric motor. In essence, an electric motor is a machine that converts electrical energy into kinetic energy by means of two interacting magnetic fields, one of which is stationary and the other of which is attached to an object capable of moving.

The genius of the Englishman Michael Faraday (1791-1867) provided the foundation for the field of electrochemistry. Faraday is often attributed with having invented the first "motor," though his simple machine was not capable of any practical application. To determine whether a current-carrying wire produced a circular magnetic field, Faraday in 1821 dangled a wire above a dish of mercury so that one end of the wire was submerged in the liquid. A small magnet floated in the center of the dish. When he ran a current through the wire, the wire circled around the

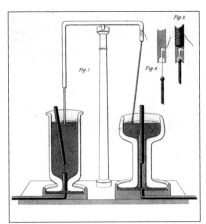

Faraday's rotating wire experiment.
(Quarterly Journal of Science, Vol XII, 1821)

magnet, thus demonstrating for the first time that electrical energy could be converted into kinetic energy.

By 1831, Henry had devised an electric motor that was far more sophisticated than Faraday's primitive device. The moving part of Henry's motor was an elongated wire-wrapped bar that was balanced atop a pyramid-shaped support in such a way that the electromagnet was free to rock back and forth on a horizontal axis. The electromagnet's polarity was reversed automatically as it rocked by means of protruding wires that connected with one of two flanking batteries. The horizontal electromagnet also was flanked at either end by two permanent magnets that stood vertically such that when the electromagnet rocked, the end pointing down would come very near

one of the vertical permanent magnets. Because of the electromagnet's changing polarity, its ends were alternately attracted and repelled by the permanent magnets, thus perpetuating the rocking movement. Henry's battery-powered machine demonstrated that continuous motion could be produced by magnetic attraction and repulsion.

Reconstruction of Henry's reciprocating motor of 1831
(Photo courtesy of the Smithsonian Institution)

THE DYNAMO

In November of 1831 Faraday went on to demonstrate magneto-electric induction, a process that yields a steady electric current. He accomplished this by attaching two wires through a sliding contact to a copper disc. When the disc was manually rotated between the poles of a horseshoe magnet, a continuous direct current was produced in the wires. Faraday's "magneto-electric machine" was the first electric generator, or "dynamo."

The same principle used in Faraday's simple generator is harnessed in today's hydroelectric generators. A hydroelectric

Faraday's magneto-electric machine of 1831
(Physique Populaire, Desbeaux Émile. 1891)

19th century Voltaic Piles

Henry's bell-ringer of 1831-32
(Smithsonian annual report for 1857)

generator uses moving water to spin a turbine which in turn rotates a series of magnets inside a generator that is attached to the turbine. The spinning magnets within the generator induce an electric current in copper coils that surround the generator. This current is delivered to homes and businesses via a grid of transmission lines.

Later, Faraday proved that the electricity induced from a magnet, voltaic electricity from a battery and the long-known static electricity were all the same thing. His discoveries contributed to the invention of the telephone, the development of the telegraph, electric lights and the production of electricity for thousands of everyday uses.

THE ELECTRIC TELEGRAPH

By the late 1800s, work in electricity and magnetism began to expand beyond the realm of pure experimentation and into the world of practical application. One of the first inventions to revolutionize communications was the electric telegraph. A predecessor to the electric telegraph was invented by Joseph Henry. In 1830, Henry demonstrated that an electric current could be sent over a distance of up

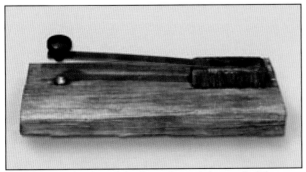

Faithful reproduction of the first telegraph key, called the "finger key "by inventor Alfred Vail.

to one mile to activate an electromagnet. When activated, the electromagnet attracted one end of a balanced lever, causing the other end to see-saw upward to strike a bell.

However, Henry did not adapt his invention to any practical use. It fell to the American inventor Samuel Morse (1791-1839) to exploit the potential of this device for practical communications. Accordingly, Morse is usually credited with the invention of the electromagnetic telegraph. In England, Charles Wheatstone and William Fothergill Cooke are credited with independently inventing the telegraph (see photo on the following page) but their system was more complicated than Morse's and did not work as well.

Both in the United States and in Europe, the telegraph had a ready market. By the 1860s it became the method of choice for transmitting information over long distances which, prior to then, could be rapidly traversed only by the Pony Express.

The basis for Morse's telegraph was that, by closing and opening a circuit, one could alternately activate and deactivate

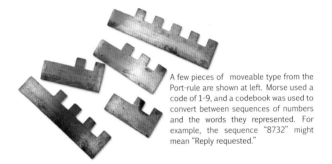

A few pieces of moveable type from the Port-rule are shown at left. Morse used a code of 1-9, and a codebook was used to convert between sequences of numbers and the words they represented. For example, the sequence "8732" might mean "Reply requested."

an electromagnet tied into the circuit at a remote location. When the electromagnet in Morse's device was activated, it triggered a magnet at the remote location that caused a suspended pencil to move, thereby making a mark on a strip of paper. Morse's stroke of genius was to modulate the pattern in which the circuit was opened and closed in a way that encoded words and phrases. The marks on the strip of paper reflected the same pattern that had been encoded into the transmitted signal. The marks on the strip could then be decoded to produce a written message, ergo the "telegram."

Morse's original code used to transmit messages was one that employed notched metal plates that had to be pulled through a slot to open and close the circuit (see photo on page 77.) However, the plates often got stuck in the slots, which inspired Morse's collaborator Alfred Vail to come up with a better way to transmit messages. Vail devised a simple switch that could be opened and closed manually to transmit the same dots and dashes that were produced by the metal plates; he also further simplified the code so that each group

Continued on page 38

Cooke-Wheatstone Double-Needle Telegraph
c. 1840s

A full size replica of first Samuel F.B. Morse's demonstration telegraph of 1837. The pen register receiver (rear) measures 29" high and 41" wide. The port rule transmitter (foreground) measures 7 x 37".

of dots and dashes represented a letter of the alphabet. Vail's agreement with Morse stipulated that Morse would receive credit for any telegraph inventions produced by Vail, so his dot-dash alphabet became known as the "Morse code."

Early telegraph transmissions were powered by battery, as no other source of electric current was available at the time. The wires were strung on poles placed next to the railroad tracks which ran across the country. Numerous relays, each providing an additional battery, had to be installed along the telegraph lines in order to boost the signal. To keep the system working, the batteries at these relays had to be frequently replaced.

THE TELEPHONE

Another significant invention based on the electromagnet was the telephone. Alexander Graham Bell (1847 - 1922), commonly attributed with inventing the telephone, reported the electrical transmission of speech for the first time in 1875. In fact, another inventor named Elisha Gray (1835 - 1901) independently invented the telephone at virtually the same time as Bell. When the two men went to court to see who had the better claim to this invention, it was Bell who prevailed. Consequently, his is the name we associate with the telephone.

Bell's invention of the telephone was an outgrowth of his efforts to improve the telegraph. Although Bell had a personal interest in inventing a way to electronically transmit sound, his financial backers were more interested in profit and instructed him to work on a new "harmonic" telegraph, that is, one that could simultaneously carry several signals. Bell accomplished this by clamping a metal organ reed between a tension screw and an electromagnet. When the reed vibrated, it touched against the screw to complete the circuit, thereby sending a series of electrical pulses at its characteristic frequency. By connecting several of these harmonic transmitters to the same wire and varying the position and thickness of each transmitter's reed, signals would be sent

Replica of Bell's 1875 "Gallows" Telephone

at different frequencies. A similar setup was attached at the receiving end where only the one reed corresponding to the frequency of the transmitting reed would respond. In this way, multiple telegraph signals could be sent simultaneously over one wire.

On June 2, 1875, while testing the harmonic telegraph, Bell's assistant Thomas Watson (1854—1934) unintentionally tightened the tension screw so much that the electrical connection was maintained continuously as the reed vibrated above the electromagnet. Bell, stationed at the other end of the test circuit, heard a clear tone instead of the expected dots and dashes. Bell immediately ran to Watson.

"What did you do?" He asked, his voice shaking with excitement. "Leave everything exactly as it is – Don't change anything – that's the current we want!"

Bell realized that the pure tone he had heard was the elusive "sound-shaped wave" that he had long dreamed about. After numerous improvements to this simple device, Bell eventually succeeded in building a contraption capable of transmitting human speech. The earliest attempt employed Bell's harmonic telegraph with only a few minor modifications. A membrane diaphragm was added, to which the metal reed was attached. The vibrating diaphragm caused a corresponding electric current - the "sound-shaped wave" that made telephone possible.

Bell's Harmonic Telegraph of 1875

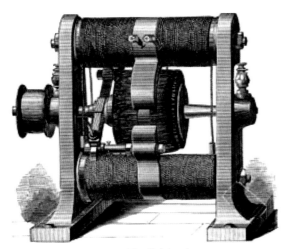

Gramme Machine lighting dynamo

ELECTRIC LIGHTING

Yet another area where electricity transformed the world is artificial lighting. In 1869, a Belgian named Zénobe-Théophile Gramme (1826 - 1901) invented a dynamo that produced direct current (DC). Gramme's dynamo was used to light towns by means of carbon arc lamps, which produce light by passing a current over a gap between two carbon rods, thereby producing a spark that emits an extremely bright light. Carbon arc lamps typically were mounted on very tall poles on the main street of a town; they were far too bright to be brought indoors. Thus, people in the late 1800s and well into the 1900s employed lamps powered by

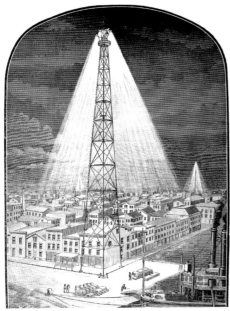

Arc lamp on tower

natural gas inside their homes and businesses. Small electric arc lights were eventually built for indoor lighting, but they were smoky, smelly, and noisy and most people stuck with gas lighting until the invention of the incandescent filament-based light bulb. Carbon arc lamps are rarely seen anymore, except for the far-reaching search lights often used to advertise public events.

Thomas Alva Edison (1847—1931) is attributed with developing the first practical electrical lights. Edison's earliest research efforts focused on improving the telegraph, and his first patent was for an improved stock ticker. The sale of the stock ticker patent financed Edison's first research laboratory, which was located in Newark, New Jersey. In 1876, Edison moved his operations to another New Jersey location, Menlo Park, there establishing an elaborate research and development laboratory from which emerged a long chain of inventions. In Menlo Park, Edison invented the first practical small motor and the tin foil phonograph. However, Edison's most significant contribution usually is considered to be his 1878 invention of a practical means of electric lighting.

Edison Electric Pen
1876

The first electric copy machine, and probably the first practical application of the electric motor.

Edison invented the electric pen as a means to produce copies of his correspondence and notes. A small electric motor operates a tiny needle in a rapid up/down motion, producing a series of small holes as the operator writes on specially treated paper. Later, a roller is used to press ink through the holes, producing duplicates of the original document.

Edison long aspired to replace the gas indoor lighting of the day with electric lighting. He perfected the first incandescent light bulb by vastly improving earlier concepts for such a device. Edison's first successful light bulb used a platinum filament. This bulb was not practical, as it burned out after only about fifteen minutes; platinum, in any case, was too costly for mass consumption. After testing numerous different materials for resistance to burning under high temperatures, Edison finally found that carbon filaments made from burned Bristol board (cardboard) worked very well (as it had already "burned") and would produce light for nearly a hundred hours.

Edison was always active in promoting his inventions. On

Continued on page 50

GRAMME MAGNETO-ELECTRIC MACHINE, No. 16. Signed "Mon Breguet.Ft. No. 836."
Mahogany base measures 13 x 11", Overall height is 22½". 42 inch large gear. Four nested permanent magnets stand 20" overall, 3" in width, 2 ½" thick. Armature measures 5" in diameter, 4" in length.

The armature consists of 30 coils, placed inside a revolving ring of soft iron. The coils are connected in series, and the junction between each pair is connected to a commutator strip on which two brushes run. The permanent magnets magnetize the soft iron ring, producing a magnetic field which induces a voltage in two of the coils on opposite sides of the armature, With thirty coils, the output is a near DC signal. A stunning example of a Gramme magneto-electric machine by one of the 19th century's top instrument makers. French, c. 1875.

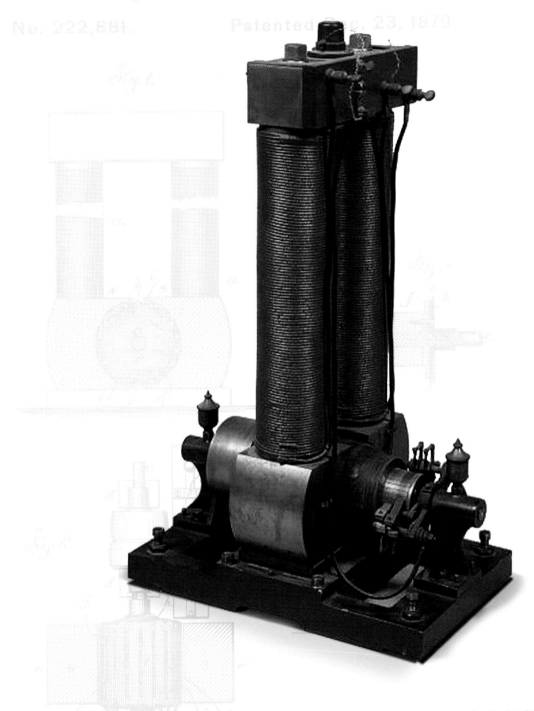

EDISON BIPOLAR DYNAMO. Unsigned. Bipolar dynamo on open-frame iron base measuring 6 x 9-5/8 x 16" tall. 2" diameter Field coils measure 9" tall. 2-1/2" dia. Armature is 4" in length, containing 32 pairs of coils. Frame is stamped "13" on one end, and "14" on the other.

A working model of the first practical dynamo for incandescent lighting. Invented in 1879 as the result of work by Edison and his chief assistants, Charles Batchelor and Francis Upton. Originally dubbed "Long legged Mary-Ann," the name was later changed to "Long waisted Mary-Ann" out of modesty. This model was obtained from the Ford Museum in Dearborn, MI. Background is the original patent illustration for the device. American, c. 1st Q, 20th century.

Electric Lighting

"I believe I can beat you making the electrical light."

- Thomas Edison to electrical inventor William Wallace, 1878

The electric light was not invented by Thomas Edison. In fact in the years preceding Edison's invention, electric light was somewhat common in the form of electric arc lamps that were in use in factories, or to light up the town square.

The light from an arc lamp is produced by passing a large current between the electrodes (usually carbon) of the lamp. The result is a very intense light, accompanied by the sounds and smells of the burning arc. While the arc lights did produce much brighter light than the gas lamps of the day, they required a huge amount of power to operate and were nearly blinding in their intensity - certainly not usable in the home or office.

By the mid-1870s, Americans William Wallace, Charles Brush, and other inventors had made small-scale arc light demonstrations. The first large-scale application occurred in March 1878 when Jablochkoff arc lamps, powered by Gramme generators, lit the streets of Paris.

These intense arc lamps were not practical for use in small locations such as the home. After visiting Wallace's factory, Edison proposed to make an incandescent lamp that was not as intense, required less power, and, unlike arc lamps, could be operated with multiple lamps in a circuit.

On December 21, 1879, Thomas Edison announced that he had achieved his goal. He produced several of the new "incandescent" lamps, and invited the public to visit his Menlo Park laboratory on New Year's Eve to see them.

The response was immediate. Several scientists proclaimed Edison's invention to be a fake. Gas stocks plummeted, and the stock of the Edison Electric Light Company soared to $3500 per share!

The demonstration consisted of about 60 lamps mounted on poles lighting the laboratory grounds and country roads in the neighborhood.   Other lamps were installed in nearby houses.   So many people came to see the demonstration that the Pennsylvania Railroad had to run special trains to accommodate them.

Today, only a handful of these first incandescent lamps survive. One of them is shown at right, and is on display in the "Electricity Sparks Invention" gallery of the Museum.

FIRST PRACTICAL INCANDESCENT ELECTRIC LAMP. Unsigned, but by Thomas Edison.   (Edison's earliest lamps feature the sunburst paper label.) Round seal stem "pantaloon" press with long platinum press leads twisted & soldered to outer leads. Screw clamps connect inner leads to filament, made of carbonized bristol board. Measures 9-1/4" in total length (including hook), 2-1/4" dia. globe, 1" dia. at neck. American, 1879

Early Edison Lamp with
Johnson bevel-ring base,
blue border label
c. 1880

Early Edison Lamp with
Johnson bevel-ring base,
c. 1881

Edison spear-point lamp with
"Petticoat" press.
c. 1880

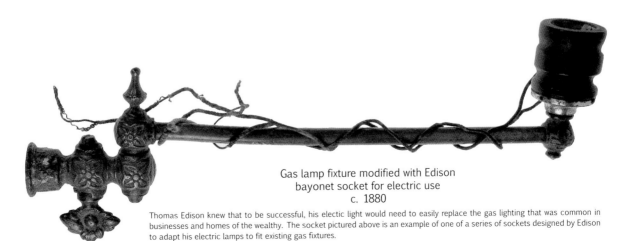

Gas lamp fixture modified with Edison
bayonet socket for electric use
c. 1880

Thomas Edison knew that to be successful, his electic light would need to easily replace the gas lighting that was common in businesses and homes of the wealthy. The socket pictured above is an example of one of a series of sockets designed by Edison to adapt his electric lamps to fit existing gas fixtures.

"This most certainly can make a bang-up socket for the lamp.."

- Thomas Edison to the "muckers," pointing out that the lid from a kerosene can would make an effective socket and base for the electric lamp.

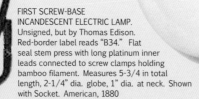

FIRST SCREW-BASE INCANDESCENT ELECTRIC LAMP. Unsigned, but by Thomas Edison. Red-border label reads "B34." Flat seal stem press with long platinum inner leads connected to screw clamps holding bamboo filament. Measures 5-3/4 in total length, 2-1/4" dia. globe, 1" dia. at neck. Shown with Socket. American, 1880

HANDLAN ST.LOUIS U.S.A.

The first experimental light sockets were made of a small wooden bowl containing two brass strips that contacted similar strips on the base of the lamp when it was inserted. A small thumb screw served to hold the lamp in place, as well as to turn the lamp off and on. Since most of the light comes out of the top of the lamp, it was desirable to install the lamp in an inverted position (hanging below the socket). But despite the thumbscrew, Edison found the lamp would eventually become loose and fall from the socket.

One night in the early part of 1880 Edison was discussing the problem with a few of his "muckers" (what he called his assistants) when he noticed a nearby can of kerosene. He immediately went over, picked up the can and examined the lid carefully, slowly working the lid back and forth on its threads. Suddenly he said:

""This most certainly can make a bang-up socket for the lamp, as well as the base!"

Thus was born the first screw-base electric lamp. One of these early experimental lamps and socket are shown above, and both are on display in the Electric Lighting Gallery at the Museum.

Unscrew any household light bulb today and you will find a standard, threaded base (known as the "Edison base".) But this wasn't always the case. In the late 19th century there were dozens of companies making electric lamps, each with its own unique base. Below are examples of a few of the more common bases sold during the 1890s.

New-Type Edison with Perkins-Mather base

New-Type Edison with Edison base

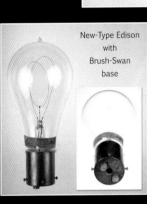

New-Type Edison with Brush-Swan base

New-Type Edison with Hawkeye base

New-Type Edison Cranberry with Westinghouse base

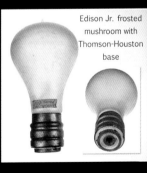

Edison Jr. frosted mushroom with Thomson-Houston base

New-Type Gilmore with Edison base

Edison with Edison base. Etching on glass reads "Edison Lab"

New-Type Edison Frosted with Thomson-Houston base

New-Type Edison with U.S. base

New-Type Edison with Combined Edison/Westinghouse base

New-Type Edison with Shaeffer base

A Few Unusual Electric Lamps

Perkins-Mather multi-filament
1880s

Perkins-Mather hairpin
carbonized paper filament
& Platinum clamps
1880s

Perkins-Mather spiral filament
1880s

Tamadine Filament, US base
1885

Swan w/ hookeye base
1880s
(Socket for screwing into gas fixtures)

Austrian deco aqua lamp
1900

Maxim, no base
1880

Swan w/ Swan base

Vic's Vapo-Lite, cobalt blue
Vic's "Vapo-rub" ointment was
placed in the cup, heat from the
lamp produced aromatic vapors.

Swan or Thomson w/ hookeye base
1880s
(Socket for screwing into gas fixtures)

New-Type Edison Color Lamps
C. 1890s

Nernst Lamp

The Nernst lamp was invented by the German scientist Walther Nernst in 1897. Instead of a filament, Nernst used a pair of small ceramic rods that were heated to incandescence. The Nernst lamp was more efficient than the carbon filament found in most electric lights of the time. The lamps were popular around the turn of the century due to their bright light, natural spectrum, and long life. However, the 1906 introduction of the tungsten filament by General Electric provided an inexpensive alternative, and the Nernst lamp fell out of favor.

Nernst Lamp with socket base
Nernst Lamp Company, Pittsburgh.
American, c. 1903

The Edison "Chemical" Power Meter

Thomas Edison set up his first power station in New York City at Pearl Street. The station began operation on September 4, 1882. In order to keep track of the electricity used by his customers, Edison designed the first electric wattmeter, called the "Chemical meter" (see below.)

It was a crude device based on Michael Faraday's principle of electroplating, Faraday had found that the transfer of metal from one plate to another in an electrolytic bath was exactly proportional to the current. Edison's first meter held a small glass jar in which two copper plates were suspended in a solution of copper sulphate. The cover of the cell was arranged so that one of the plates was easily removable by means of an insulated clamp with a thumb screw; the other plate, which was thick and cylindrical in form, was intended to remain in the cell to allow the copper to be transferred from it to the other plate via electrolysis. An Edison employee would visit the meter periodically, remove the electrode and weigh it, and the customer would be billed accordingly.

While the concept worked well in theory, in practice it was inconvenient and not especially accurate. Worse, there was no way for a customer to independently confirm their consumption of electricity so confidence in the device was not high. As a result the meters were replaced in short order, and very few can be found today.

Inside the Edison Chemical Meter.
Note the small jar containing the electrodes and electrolyte solution.

MAZDA Electric Lamps

The trademark MAZDA was not the name of a lamp but rather the mark of a service. The word MAZDA was first used on lamps on Dec 21, 1909. MAZDA lamps made from that date to about May of 1911 utilized sintered (made by mixing metal powder with a filler, then formed into a wire) tungsten filaments, whereas those made after May 1911 utilized drawn tungsten filaments Westinghouse used the name MAZDA starting in 1912. The name was not used by the General Electric Company after 1945.

MAZDA LAMP POINT-OF-SALE DISPLAY. Signed "General Electric Co." Walnut base with painted metal display measures 27$^{1/4}$ x 5$^{7/8}$ x 13" tall at highest point (not including lamps). 11 Edison sockets are spaced evenly along the outside edge of the convex display, each with a SPST power switch, controlling a tipped Edison base lamp. The lamps are not Mazda Tungsten lamps, rather they are carbon, double loop, anchored, c. 1910. Design and graphics by noted artist Maxfield Parrish. This apparatus was used as a point of sale device for displaying Mazda lamps as well as an in-store test fixture for customers to test their lamps before purchase. American, c. 1920s

Thomas Edison with "Jumbo"

A means of sending electricity over longer distances was sorely needed. The solution to this problem proved to be the invention by Nikola Tesla (1856—1943) of a system of motors and generators that operated on alternating current (AC). Tesla, an immigrant from Croatia, worked for Edison for a period after first arriving in this country. However, the two men did not get along. Tesla left his employment with Edison and went to work for George Westinghouse, an early competitor of Edison who hoped to break into the electric lighting business. Westinghouse immediately saw the value of Tesla's AC motor. He began building AC generating plants, which could send their power over much greater distances than Edison's generators. Westinghouse and Tesla soon became a threat to Edison's business plan.

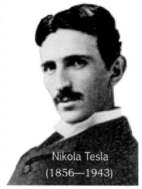

Nikola Tesla (1856—1943)

December 31, 1879, he arranged an outdoor demonstration of his electric lighting technology at his Menlo Park laboratory. The demonstration was highly successful in forging a link in the public eye between Edison's name and electric lighting. Although Edison required another year to perfect his new technology, the Menlo Park demonstration primed his contemporaries to rapidly embrace the new electric lighting as soon as it hit the market. One of the few remaining Menlo Park demonstration light bulbs is on display at the Museum.

To power his system of lighting, Edison designed an improved dynamo, known as "Jumbo," and, in 1882, opened the first commercial power station on Pearl Street in New York City. Using six Jumbo dynamos, the Pearl Street plant was capable of producing 100 kilowatts of direct current, The amount of electricity needed to light up the one square mile serviced by the plant. Busi-

A long battle ensued in which Edison tried to convince the public that Westinghouse's AC was far too dangerous for everyday use. This, ironically, was the same tactic that the gas companies had used previously in efforts to scare the public away from Edison's own DC technology. Again, the superior product won out. The Edison Light Company was eventually bought out by a rival company, which later became the General Electric Company (GE). Edison was devastated that his name was dropped from the name of the company that bought him out.

Thus, by the end of the nineteenth century, the theory of electrostatic charges had developed into the basis for power generation via electromagnetic generators. The new technology had made it possible for people to light their homes without kerosene

U.S. DEPARTMENT OF THE INTERIOR, NATIONAL PARK SERVICE, EDISON NATIONAL HISTORIC SITE

Thomas Edison with his team at Menlo Park, New Jersey. Spring of 1880. Edison is seated in middle, second row from the top.

nesses in the Pearl Street district began converting to electric lighting despite dire warnings from the gas companies about the dangers of proximity to electricity. However, Edison's success was limited by the fact the range of his DC generator was only about one mile.

or gas. Electrically-powered communication by telegraph and telephone profoundly changed the scale of distance between people. The age of electric machinery, pioneered by Thomas Alva Edison, was on its way.

Perpetual Motion

IN THE BEGINNING of the 19th century, the abbot Zamboni, an Italian physicist, developed a method of making Voltaic piles (early batteries) by using very thin metal foil and paper. This method enabled Zamboni to create piles of over 2000 layers that stood less than 12" tall.

Producing over 2,000 volts, batteries such as these created an electrostatic charge on their terminals. Taking advantage of this, Zamboni manufactured an instrument composed of two Voltaic piles that alternatively attracted or repelled a pendulum situated between them. The mechanism formed the basis for the first electrostatic clocks.

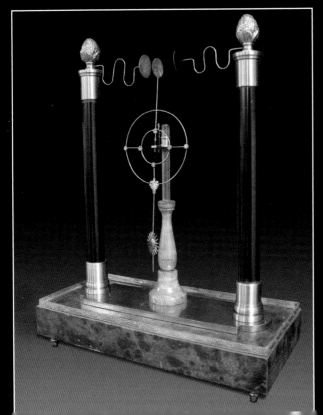

Shown here are two of the Zamboni perpetual motion machines that are on display at the museum. Although the piles are by now exhausted, Some say "perpetual motion" machines such as these worked non-stop for more than 100 years.

These very rare pieces represent the earliest research on very high voltage batteries; today only a few examples survive.

A Slight Digression:
Entertainment in the Home

OVER A CENTURY BEFORE the dawn of the phonograph and radio, more people owned music boxes than any other type of musical instrument. Besides mastering a musical instrument or singing, music boxes were the only way people could enjoy music at home.

Music boxes are powered by hand cranks or by wind-up spring motors, and they all employ the same principle: they pluck a tuned comb to make music. The music is "stored" on pinned brass cylinders, punched metal discs, or even paper rolls. The Museum's collection includes a music box of the metal disc variety, manufactured by the Criterion Company. This music box was given to the mother of Curator Jonathan Winter by the Chief of the San Francisco Fire Department in 1905, just one year before the great earthquake and resulting fires razed San Francisco.

Because the metal discs could be stamped out at a factory, this form of music box offered the least expensive and most varied source of music for home entertainment. The Museum has produced a CD of all twenty eight of the Museum's Criterion Music Box Songs. This CD is available in the Museum store.

"Nipper"
Radio Corporation of America
c. 1920s

In 1877, Edison's phonograph was introduced and made possible the recording and playing of music at home using wax cylinders. Although these early machines were quite expensive, much cheaper and affordable wind-up gramophone/phonograph machines soon became available and sales soared. By the time RCA and others began to mass produce the radio for home use in the 1920s, there already was a mass-market ready, willing and able to buy.

c. 1900 Home Entertainment

Edison Home Phonograph
with Morning Glory Horn
c. 1908

Columbia Disc Phonograph
c. 1908

Criterion Music Box
c. 1908

Edison Tin-foil Phonograph
Replica

Edison Home Phonograph
c. 1908

Graphophone Model Q
Columbia Phonograph Co.
c. 1900

Holmes Stereoscope Viewer

Holmes Stereoscope Viewer

Edison Wax Cylinders
c. 1908

Electric Motors

Defining "motor" as any apparatus that converts electrical energy into motion, most sources cite Michael Faraday as developing the first electric motors in 1821 (See page 56, bottom right.) They were useful as demonstration devices, but had no practical application, and most people wouldn't recognize them as anything resembling a modern electric motor. During this time a number of innovative ideas emerged using various combinations of electromagnets and approaches for maintaining continuous mechanical motion - but electric motors remained little other than curiosities.

The role of the electric motor (or "electro-magnetic engine," as it was called) as a tool for real work did not emerge until the 1840s. These motors were characterized by their close resemblance to steam engines. A fundamental limitation remained however; there was no source of electrical energy sufficient to supply the engines with enough power for them to compete with steam engines. So electric motors were relegated to tasks such as desk fans, etc., which required little power.

Surprisingly, it was the electric light that paved the way for modern electric motors. In order for electric light to be practical, a new power source, something other than batteries, had to be found. The development of the electric dynamo (generator) in the 1870s not only enabled electric lighting, it also provided a powerful source of electricity to run large electric motors, ending the era of the steam engine.

The Museum contains one of the largest collections of early electric motors in the United States.

ELECTROMAGNETIC ENGINE, by Gustav Froment (unsigned). Missing wooden base, the cast iron open frame measures 12-1/4" x 4-1/2" and is 6-1/4" high including magnetic armature. Four large electromagnets measuring 2-1/8" in length and 1-1/2" diameter. French, c. 1848.

DEMONSTRATION ELECTRO-MOTOR TRAIN, made of brass, steel, copper and wood, measures 5-1/2" length. Early example of an electromotive train. Has two electromagnets, a 10 pole rotor geared to the main axle, the other axle free to pivot. Designed to ride on a 2-1/2" gauge track. French, c. 1850-1860.

TESLA POLYPHASE MOTOR. Signed "ARRIGHINI ANGELO COSTRUTTORE, MILANO." Made of brass, wood, copper, steel, iron and ebonite. Four electromagnets are supported on a 9 1/4" circular iron core that reaches 18" at its highest point. The 4-pole rotator is hinged, allowing the motor to operate in a vertical or horizontal plane. The mahogany base measures 22 1/8 X 12", supporting 6 brass binding posts and a pole reversing switch. Italian, c. 1900.

Electric Motors

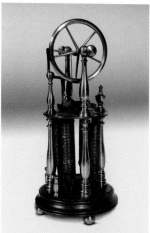

UPRIGHT RECIPROCATING ENGINE
Unsigned by apparently constructed by Daniel Davis, Boston. 5" diameter walnut base supported by four brass ball feet; 11-1/2" height, American, c. 1842.

ELECTRODYNAMIC REVOLVING RING
Unsigned but apparently by Daniel Davis. Boston. 5" diameter rosewood base. American, c. 1842.

APPARATUS TO EXHIBIT THE ROTATION OF TWO HOLLOW METALLIC CYLINDERS ABOUT THE POLES OF A MAGNET WHEN AN ELECTRIC CURRENT FLOWS THROUGH THE CYLINDERS
Unsigned, but pictured in Watkins and Hill Catalog of 1845 on page 8, number 67. Made of iron, steel, ebony and ivory. This apparatus in 9" at its highest point. Likely English or Italian, c. 1840.

REVOLVING OR "ROTATING" BELL ENGINE.
Unsigned, but apparently by Daniel Davis, Boston. 5" diameter walnut base; height, 13-1/2". American, c. 1842.

REVOLVING SPUR-WHEEL.
Unsigned but apparently by Daniel Davis. Boston, 4-1/2" diameter walnut base, 13-1/4" in height, American, c. 1848.

EARLY FARADAY ROTATING WIRE EXPERIMENT. One of the first electric motors. . A permanent magnet is placed vertically inside a container of mercury. A copper wire is suspended into the mercury from above. An electric current is passed through the wire and mercury, creating and electromagnetic field around the wire, which interacts with that of the permanent magnet, causing the wire to rotate around the magnet. c. early 1830s

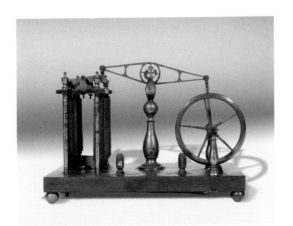

MAGNETIC BEAM ENGINE. Signed "Pike and Son, New York." 10"x5" walnut base; base supported by four brass ball feet; 7" height. American, c. 1840.

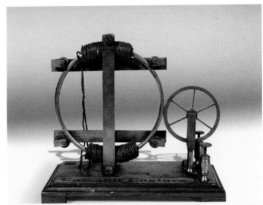

PAGE'S AXIAL REVOLVING CIRCLE. Unsigned but probably by Daniel Davis or Charles Page. Bimetallic ring gear mounted on walnut base 9 " x 4-1/2"; 8" height. American, c. 1848.

RITCHIE'S MOTOR. Very primitive electromagnetic motor invented by Rev. William Ritchie; "probably the first man to produce the rotary motion of an electromagnet," in 1833 Walnut base measures 6-1/2" diameter. American, c. 1830s

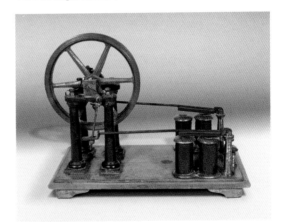

ELECTRO-MAGNETIC MOTIVE MACHINE. Unsigned but likely by Watkins and Hill. Mahogany base measures 14" x 7"; with brass wheel height is 11". American, c. 1845.

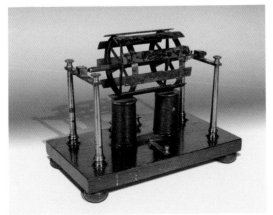

MAGNETIC MOTOR, made of wood, iron, brass, steel and copper. Mahogany base measures 12" x 7" and stands 10" in height. The wooden base being one inch in thickness and supported by four wooden bun feet. The large nine-pole rotator is supported by four turned brass columns. There are two large vertically positioned electromagnets measuring 3" in height and 1-3/4" diameter. Probably English c. 1850.

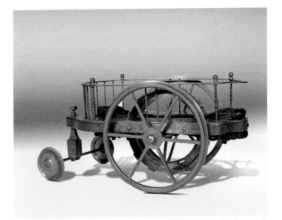

DEMONSTRATION ELECTRO-MOTOR TRAIN. Unsigned. An early example of an electromotive "train." It is 8" long, equipped with four electromagnets, a twelve pole rotor on the main axle, commutator wheels and contacts. The six-spoked main wheels are designed to ride on 3-3/4" gauge track; the front wheels are off center; designed for a circular track of fixed radius. French, c. mid- 19th century.

Induction Coils

The principle of electromagnetic induction was discovered by Michael Faraday (1791-1867) in 1831. Induction coils were used widely in electrical experiments and for medical therapy during the last half of the 19th century, and their use eventually lead to the development of radio in the 1890s.

An induction coil essentially operates like a big buzzer (See diagram.) The interrupter's contacts (C), are wired in series between the battery (B) and the coil. The contacts are closed in their resting state. When key (K) is closed, current flows from the battery (B) through the coil, producing a magnetic field in the iron core of the coil. The magnetic contact arm (V) is pulled toward the core, opening the interrupter contacts and breaking the circuit. The falling magnetic field relaxes its hold, the contacts close, and the entire cycle repeats for as long as the key is pressed.

The rising and falling magnetic field of the primary coil induces a high voltage across the secondary winding.

Induction coils were used in many different kinds of experiments. In order to facilitate easy access to a variety of devices, an experiment table was commonly used. The output of the coil was connected to vertical brass uprights and the connections could be adjusted safely using insulated handles.

Induction coils were also used in experiments or treatments where the output of the coil was connected to a part of the body. Many people of the time believed the mild electrical shock to be therapeutic. An entire industry of medical devices was born from the invention of the induction coil. While most of these fall into the category of "quack medicine, " some applications proved legitimate. For example, electro-stimulation is still in use today in physical therapy.

A specialized type of induction coil called a Rhumkorff coil was used in early radio transmitters - the so-called "Spark" transmitters of the early 20th century.

COMPOUND MAGNET AND ELECTROTOME. Unsigned but by Daniel Davis. On walnut base measuring 9" x 4" with 4 brass ball feet. Invented by Charles Page in 1838 but patented in 1868 (see Page's biography: "Physics, Patents and Politics" by R.C. Post, p. 24.) First described by Page in "Magneto-Electric and Electro-Magnetic Apparatus and Experiments," American Journal of Science, (1839), pp. 252-68, figs. 1-7. American, c. 1840.

PIKE'S ROTARY MAGNETIC MACHINE. Signed "Benjamin Pike, Jr., " an electromagnetic machine and battery with trade card attached to inside cover of walnut box measuring 16" x 6" x 7". Trade card reads "Benjamin Pike, Jr. Optician, Importer and Manufacturer of Mathematic and Philosophical Instruments, No. 294 Broadway, New York". Elaborate label on inside top lid of box reads "Pike's Rotary Magnetic Machine introduced Nov. 1843" with extensive instructions on use. The kit contains a electromagnetic medical device and battery all in fine condition with some loss of paint on battery. The device is on a walnut base with brass ball feet and measures 10" x 4". The battery measures 4-7/8 " x 8 " in diameter and 5-1/4 " in height. This machine is described in Pike's Catalog Vol. II on pages 44- 47 and pictured in Fig. 468. The device is featured on the front wall (in a large advertisement) of Pike's first shop in 1848 as seen in the woodcut frontispiece in his catalogs of 1848 and 1856. American, c. 1848.

MOORHEAD'S
IMPROVED GRADUATED

MANUFACTURED AND SOLD

R MAJOR Sc

WHOLESALE AND RETAIL BY

117 FULTON ST.

No 182
D C. MOORHEAD BROADWAY NEW YORK.

Entered according to Act of Congress, in the year 1848, by D. C. Moorhead, in the Clerk's Office of the District Court of the Southern District of New-York.

Electro-Therapeutics

"I need hardly recall to mind, that until quite recently, to venture to speak of electricity as a curative power was pretty certain to result in the speaker being branded as little better than a quack."

- Herbert Tibbets, M.D., A handbook of medical and surgical electricity. 1877

The use of electricity's magical properties for treating the various ills and infirmities of the human body dates back to the Roman empire and is probably the first practical application of electricity. As far back as 42 BC, Roman physicians treated gout by placing the poor victim's affected limbs into water occupied by electric fish. Pliny the elder (AD 23 - 79) described necklaces of amber worn by women and children for the benefit of their supposed curative powers.

Centuries later, the invention of the static electric machine would enable broader use. In 1744, A German physician named Kratzenstein recorded a case of cure of paralysis of the fingers by sparks drawn from a frictional static electricity machine. The first book dedicated to medical use of electricity ("Experiences sur Electricite") was published by Jean Louis Jallabert of Geneva in 1749, in which he reported a cure of paralyses of the right arm by application of electric sparks. Some scholars regard this as the first documented "proof" that paralysis could be successfully treated by electricity.

In 1755, Benjamin Franklin, never one to shy away from experiments on his fellow man, reported an experiment on electrically induced amnesia: *"I laid one end of my discharging rod upon the head of the first; he laid his hand on the head of the second; the second his hand on the head of the third, and so to the last, who held, in his hand, the chain that was connected to the outside of the jarrs. When they were thus placed, I applied the other end my rod to the prime-conductor, and they all dropt together. When they got up, they all declared they had not felt any stroke, and wondered how they came to fall; nor did any of them hear the crack, or see the light of it."*

The invention of the electric battery, induction coil, and magneto in the early 19th century provided new ways to produce the mysterious and powerful effects of electrical stimulation, and by the mid 19th century, Electro-Therapy was a generally accepted form of medical treatment. The field was divided into three major sub-categories:

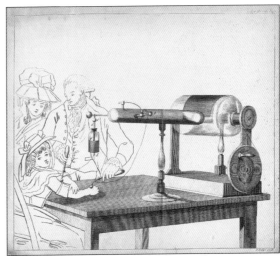

Early use of "Franklinic" Electro-Therapy
An essay on Electricity. Adams, George - 1787

Franklinic (static electricity), Galvanic (direct current), and Faradic (high voltage from induction coils or magnetos)

"Each of the three kinds of electric currents—galvanic, faradic, or franklinic— exerts a favorable influence if employed for a certain period. Franklinisation improves most rapidly the symptoms of irritation, especially the trouble of sensation and coordination. But for the same period of treatment the results acquired by constant currents are without doubt the most lasting. Induced (Faradic) currents hold an intermediate position between the other two varieties of current..." "....At the first sitting he gives a successive application both of both the constant and induced currents. Simultaneous application of both the faradic and constant currents is used at the next sitting. At the first sitting, it is sometimes useful to substitute the static electricity for the induced current."

. Journal of the Medical Association of New Zealand. Vol III, 1889-90

Man receiving "Faradic" Electro-Therapy
Electricity in Medicine. Ambrose L. Ranney, M.D.   D. Appleton & Co. 1885 p. 73

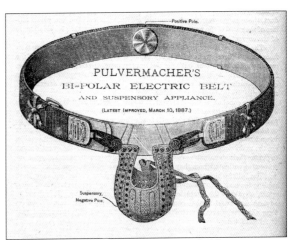

"Galvanic Bi-Polar Electric Belt"
Source: Pulvermacher Company catalog

Standing nearly 6 feet tall, This impressive Holtz Electrostatic generator is a beautiful example of "Franklinic" apparatus from the turn of the century. The machine could have been used to power X-Ray or ultra-violet therapy tubes, or to apply electricity directly to various places on a patient's body, as shown in the 1904 illustration below.

HOLTZ PATTERN ELECTROSTATIC GENERATOR. Signed "Van Houghten & Tenbroeck Co. New York". Oak frame measures $65^{1/2}$ x $33^{1/2}$ x $69^{3/4}$" tall, supported by four curved oak Queen Anne legs with claw foot detail. Contains 12 ea. 32" glass disks. Fine condition noting reproduction Leyden jars. American, c. 4th Qtr, 19th Century.

Source: "Electricity in Everyday Life". Vol III, Houston, Edwin J. P.F. Collier & Son, NY. 1904.

EARLY PHOTOTHERAPY APPARATUS. Signed "E. BALZARINI, MILANO". Brass frame measures $9^{3/4}$ dia. x $18^{3/4}$" tall, supporting four horizontal $3^{3/4}$" long brass cylinders, each containing a miniature Edison lamp socket, whose wiring is brought to a central terminal at the top of the device. The four Edison sockets contain lamps of green, yellow, red, and blue, respectively. Italian, c. 4th Qtr, 19th Century.

D'ARSONVAL SPIRAL HELMET. Signed "E. BALZARINI, MILANO". Brass frame measures $9^{3/4}$ dia. x 21" tall, supporting a standard brass Edison lamp socket at the top of the device. Italian, c. 4th Qtr, 19th Century.

ELECTRIC PHRENOLOGY APPARATUS. Consists of two parts, a wood box containing a sledge induction coil and three batteries, and a headpiece. The headpiece is of 5/8" wide brass strapping, forming a crown 9" in diameter and $5^{1/4}$" tall. 13 brass electrodes, each 3" in length, are evenly spaced across the crown. Signed "E. BALZARINI, MILANO". The induction coil measures 3" dia. x $3^{3/4}$" in length, mounted on a 14" sledge track. Also three dry-cell batteries, each measuring $2^{1/2}$" dia x $5^{1/4}$" tall. Coil and batteries are contained in a soft wood box measuring $19^{3/4}$ x $11^{3/4}$ x $7^{1/4}$". Plate inside box lid reads: "IMPIANTE ELETTRICI, MARINELLI, CAV. UMBERTO MARINELLI, L' AQUILA". Italian, c. 4th Qtr, 19th Century.

ELECTRICAL BELT, signed "Pulvermacher Galvano Co. Electricity is Life" in black cloth covered box measuring 8-5/8" x 61-3/8" (h) and accompanied with instruction book and catalog by the Pulvermacher Galvano Company. Belt is made of a brass weave. Fine condition. American. c. 1880s.

MAGNETO-ELECTRIC MACHINE. Signed "Alfred Apps, Maker. 433 Strand, London." Mahogany case measures $6^{3/8}$ x $6^{1/8}$ x $7^{3/4}$". Beautifully constructed magneto electric generator mounted to mahogany base measuring $5^{1/2}$ x $5^{1/2}$ x 3/4", standing 7" tall. Includes two hollow brass handles $3^{3/8}$ x 7/8" dia. British, c. 4th Qtr, 19th Century.

ELECTRIC BRUSH. Signed "LOUIS BRANDUS, Boulevard Bonne Nouvelle 35, PARIS". Original box. Gutta purcha handle measures 5 x $2^{3/4}$" and supports brush of fine copper wires. Hollow interior holds 4 zinc and 4 copper plates, plus dry sponge. c. 4th Qtr, 19th Century.

ELECTRO-MAGNETIC HAIRBRUSH quack medical device used to stimulate hairgrowth. Made of gutta purcha, brass and nickel plated brass. Signed "A. Gendron, Bordeaux" and stamped "F. 72". The eight sided gutta purcha handle is 4-1/2" in length and leads into finely turned brass column (3-5/8 " length) which is attached to the nickel plated brass plate of numerous brass spikes, the plate is 3" in diameter. French, c. 1870s. Very fine condition.

PORTABLE X-RAY AND HIGH-FREQUENCY COIL WITH FLUOROSCOPE. Signed "Campbell Bros. Lynn, Mass. USA". Oak suitcase measures 22 x 10 x 8". Height to top of X-Ray tube when assembled is 33". Base of case contains high-voltage coil used to power the X-Ray tube or one of an assortment of ultra-violet therapy tubes that are mounted in clips on the inside top cover. American, c. 1920.

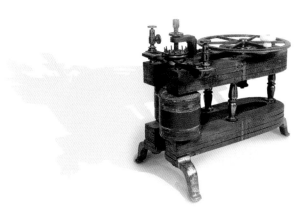

IMPROVED MAGNETO-ELECTRIC MACHINE, unsigned but by Daniel Davis. 13-1/2" x 7" with 8 painted U-shaped magnets supported on Queen Anne legs; 9" height; with walnut box 16-1/2" x 9-1/4 x 12", Also contained are two rosewood handle accessories for medical-electric use. Described in Manual of Magnetism p. 277 Fig. 174 (2nd ed). Fine condition noting minimal wear to magnets. c. 1848.

"DR. S.B. SMITH'S 18 POUND MAGNETIC CURRENT AND 6 POUND DIRECT CURRENT MACHINE", signed in gold lettering as written, also with " No. 309 Broadway." Box made of walnut measuring 6" x 3-1/2 " with electromagnetic machine and accessories. Label of inside cover reads "Dr. S.B. Smith's Pocket Magnetic machine of very strong Intensity" with instructions. Fine condition. c. 1845.

IMPROVED MAGNETO-ELECTRIC MACHINE signed on inner lid of mahogany box measuring 11-1/4" x 5-3/4" "W.C. & J. Neff, Philadelphia. No. 3 1/2 South Seventh Street;" note two (painted red and green) horseshoe magnets with red coiled electromagnets. Very fine condition. Neff was at this address in the 1850s. c. 1850-1860.

DAVIS AND KIDDER'S PATENT MAGNETO-ELECTRIC MACHINE signed and patented by Ariel Davis in 1854 (brother of Daniel Davis.) An early example of this machine, signed on paper label inside lid as well as on interior brass frame and on top of mahogany wooden box. Box measures 9-3/4" x 4-1/4" x 4-1/2" with beveled brass edges. Machine is complete with early rosewood and brass handles; horseshoe magnet is painted red and the electromagnets are covered in red felt; all indicators of early manufacture. Fine condition, American 1854.

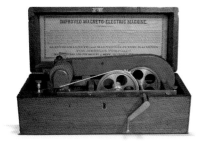

IMPROVED MAGNETO-ELECTRIC MACHINE signed "W. C and J. Neff, Philadelphia. 11-3/4" x 5" mahogany box; with one green painted horseshoe magnet and red coiled electromagnets. Very fine condition. Neff was at this address in the 1860s and 1870s. c. 1870.

ELECTRO-MEDICAL INDUCTION COIL, signed "Ruhmkorff Mecanicien, 15 Rue des Macons Sorbonne, Paris" on ivory label. Mahogany box measuring 10-1/2" x 4-1/2" x 2" See pages 250-251 of The Practice of Science in the Nineteenth Century by Gerard L'E Turner. Very fine condition noting missing small glass bottle. c. 1850.

Electrical Measurement

Simple voltmeters were developed in the 18th century (see *Electroscopes and Electrometers* on page 27), but galvanometers were the first instruments used to determine the presence, direction, and strength of an electric current in a conductor.

Projecting Electroscope with Grenet Cell
c. 4th Q, 19th Century

All galvanometers are based upon the discovery by Hans C. Oersted that a magnetic needle is deflected by the presence of an electric current in a nearby conductor. When an electric current passes through a conductor, the magnetic needle tends to turn at right angles to the conductor so that its direction is parallel to the lines of induction around the conductor. The north pole of the needle points to the direction in which these lines of induction flow. The extent to which the needle turns depends on the strength of the current.

André-Marie Ampère, (1775-1836), is credited with inventing the galvanometer in 1824. The earliest galvanometers were literally constructed of a compass surrounded by a coil of wire (see example below.) These meters were called "sine" galvanometers because the sine of the angle of deflection of

the needle is proportional to the strength of the current in the coil. At this point in time it was impossible to construct a meter whose needle deflection was directly proportional to the current under measurement.

Unfortunately, simple galvanometers such as the Struers model shown below (left) were inaccurate and inconsistent in their readings. By placing the compass at the center of a precisely calculated circle, accuracy could be improved substantially (below). Other improvements were added later, including replacing the compass with a specially designed meter movement or adding leveling screws.

These large stationary-coil type galvanometers were used as the standard current measuring instrument into the last quarter of the 19th century.

Tangent Galvanometer
Philip Harris Ltd, Birmingham, England
c. 1st Q, 20th Century

Reflecting Galvanometers

One of the limitations of early galvanometers was that the length of the needle had to be kept very short in order to minimize the effects of the earth's magnetic field and reduce damping errors introduced by the mass of the needle itself. Unfortunately, the shorter the needle, the less distance the tip will travel as it inscribes an arc, and thus the more difficult it will be to read very small changes in current. This problem was solved ingeniously by using a beam of light as the needle; a shaft was placed through the center of the needle and a very small mirror was attached. A beam of light is reflected off of the mirror and onto a scale located about three feet away. The result is that an extremely small deflection of the mirror will cause a much larger movement of the beam on the scale.

Struers Sine Galvanometer
c. 1st Q, 20th Century

Reflecting Galvanometers

Reflecting Galvanometer
F.E. Becker & Co.
American, c. 1910

Mirrored Galvanometer
English, c. 1900

Reflecting Galvanometer Scale
French, c. 4th Q, 19th Century

Reflecting Galvanometer
Knott Apparatus Company
American, c. 1st Q, 20th Century

Moving-Magnet Reflecting Galvanometer
Cambridge Instrument Co. Ltd.
British, c. 1905

Hughes Telegraph
Signed "Dumoulin - Froment a Paris"
French, Mid 19th Century

Operated on the Paris - Milan line

The Telegraph

The word telegraph comes from the Greek words for "writing" (graphe) and "far" (tele). Literally, "telegraph" means "writing at a distance."

Most people think of Morse code when they hear the word telegraph. But technically speaking, telegraphs go all the way back to ancient times. The signal fires of the ancient Greeks, smoke signals of the American Indians, even the famous lantern in Boston's North Church ("One if by land, two if by sea...") are all examples of early telegraphs.

These early telegraphs were called optical or visual telegraphs. Electric telegraphs came later, and culminated finally with the electromagnetic telegraph, for which Samuel Morse is famous. A lesser known electromagnetic telegraph is the Hughes model shown at left, which was operated by pressing keys on a piano-style keyboard!

The Museum contains one of the largest collections of early telegraph apparatus in the country.

Edison
Universal Stock Ticker
American, c. 1870s

Optical Telegraphs

In 1791, Claude Chappe invented a telegraph system that used large wooden paddles attached to towers.   Using a telescope, a spotter standing in one tower could see the signal from the previous tower.   The spotter would set his paddles to match and the message would travel in this fashion from tower to tower.   By 1794, the Chappe system made it possible for the French to send a message from Paris to Lille - a distance of some 191 kilometres - in 15 minutes. Realizing the strategic advantage of rapid communication, Napoleon had the Chappe telegraph installed on towers all across Europe. It is said that a message could be sent from Paris to Milan in as little as six hours!

Although effective, the visual telegraphs had one big limitation: They didn't work at night, or in heavy fog.

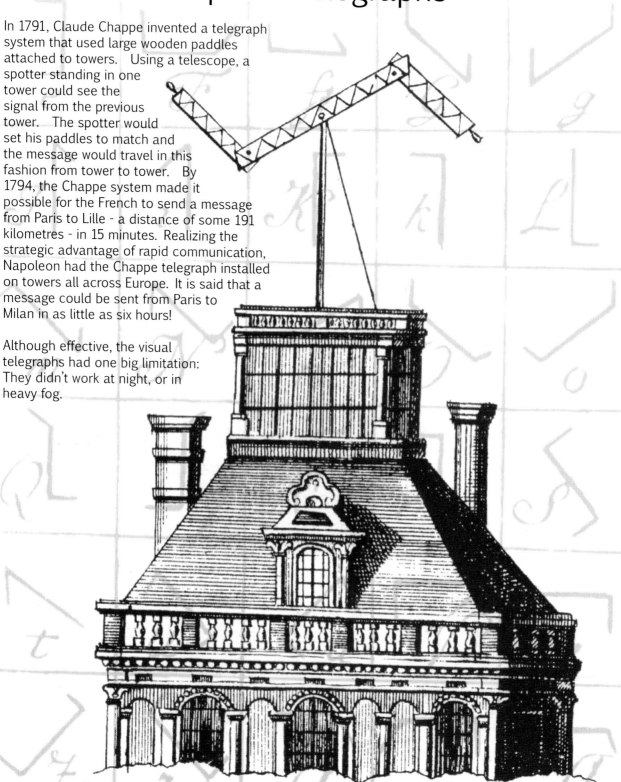

Chappe Telegraph installed at the Louvre, Paris

Source: "Beschreibung und Abbildung des Telegraphen," Leipzig, 1795, pl 1

Electric Telegraphs

The first electric telegraphs were made in the early 18th century using static electricity. After the invention of the battery in 1800, more practical designs for the telegraph were suggested, taking advantage of the continuous electric current supplied by the battery. It wasn't until Oersted's 1821 discovery of electromagnetism and the development of the electromagnet by Sturgeon and Henry ten years later, that a flurry of study, experimentation, and invention took place and the idea of an electromagnetic telegraph was born.

A number of novel ideas were put forth and though some were technically workable, the public wasn't ready at that time to accept the idea of communicating over wires. Many considered it quackery or magic.

Soemmering Electric Telegraph
1809

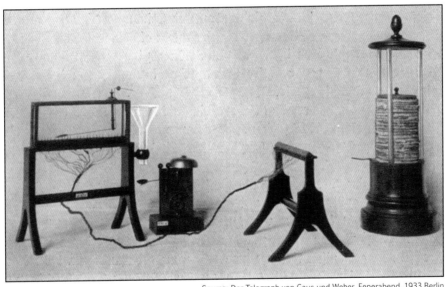

Source: Der Telegraph von Gaus und Weber, Fenerabend, 1933 Berlin

Needle Telegraphs

Prior to the popularization of the electric telegraph, the most well known visual telegraph was the Chappe system, which extended throughout Europe (see page 70.) Railway semaphore signals were also in common use. So it was natural that the first electric telegraphs would mimic these systems, using needles in place of large wooden paddles or semaphore flags. Eventually the Wheatstone & Cooke system dominated as the most popular general purpose needle telegraph, although semaphore needle telegraphs were still in common use by the railroads into the late nineteenth century. Below are some of the needle telegraphs on display at the Museum.

THE COOKE - WHEATSTONE TELEGRAPH

If you grew up in the United States, you were likely taught that the telegraph was invented by Samuel Morse. However, if you went to school in Great Britain, you probably learned that Charles Wheatstone and William F. Cooke were there first.

Shown here are beautiful examples of Cooke and Wheatstone needle telegraphs. The two-needle version is very rare, one of a handful in the United States.

The first Cooke and Wheatstone telegraph was patented in 1837, the same year Morse received a patent for his telegraph. It was installed in London as the world's first public telegraph service in 1839, running 13 miles from Paddington to West Drayton.

The electric telegraph found an ideal market in the developing railway network, which was the first practical use of electricity for long-distance communication.

Cooke-Wheatstone
Double-Needle Telegraph
c. 1840s

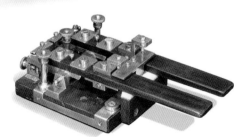

Double Current Telegraph Key
British Post Office
c. 1860s
Used with the Cooke-Wheatstone Single Needle
Telegraph

Cooke-Wheatstone
Single-Needle Telegraph
c. 1850s

ABC and Dial Telegraphs

While the Morse system was the only telegraph technology to achieve success in the U.S., Other designs found success for a limited time in Europe. One of these was the ABC telegraph, along with its cousin, the dial telegraph. In both types the transmitter consists of a large pointer, similar to the hands on a clock, surrounded by the characters of the alphabet, similar to the numbers on a clock face. The operator sends a message by rotating the large pointer from letter to letter, spelling out the message. On the receiving end, a similar device using a smaller needle as a pointer follows along, pointing out the characters of the message as they are received. This sounds like a much simpler method than Morse since no codes are involved, but in reality, the system was fraught with problems. The mechanism was complex and subject to jamming or breakdown, and the need for synchronization between the transmitter and receiver meant that messages were frequently received garbled or incomplete.

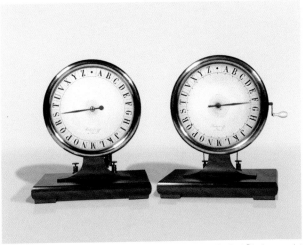

RUHMKORFF ALPHABET DIAL TELEGRAPH. Signed "Ruhmkorff a Paris" in fancy script lettering on silvered background showing the letters of the alphabet. Circular dial being 8-3/8" in diameter, mounted perpendicular to mahogany base 9-1/2" x 6-3/4". This telegraph is an early example of the Wheatstone apparatus and is made from silver, brass, mahogany and iron. French, c. 1840.

ALPHABET DIAL TELEGRAPH. Signed "Breton Freres, Paris, Rue Dauphine, 23." A pair of combination sender/receivers made from walnut, brass and steel. The base is 8" x 7" and the apparatus is 7-3/4" in height. French, c. 1860.

DIAL TELEGRAPH. Signed "Deleuil A Paris" on walnut base measuring 12 x 8". The brass dial, containing the letters of the alphabet and numbers 1 through 9 engraved on its face, stands 7-3/4" at its highest point; the dial itself is 5-1/2" in diameter. Constructed of wood, brass, steel, ivory with silk-covered green wire electromagnets which comprise the sounder/receiver. French, c. 1855-1865.

BREGUET ALPHABET DIAL TELEGRAPH. Sender and receiver constructed of mahogany, ebony, brass and steel. Receiver measures 8-1/2" x 5-1/2"; height is 7-1/2". Sender measures 8" x 7-1/2" and height is 4-3/4" (including handle). French, c. 1860s.

Printing Telegraphs

Probably the most technologically impressive of the early telegraphs were the printing telegraphs of Hughes, House, and others. These machines shared one advantage with the ABC telegraph – no codes were used. Instead, the operator at the sending station sat at a keyboard that resembled a piano, right down to the black and white keys. Each key on the "piano" represented a character of the alphabet which, when pressed, caused the corresponding character to be printed on a strip of paper at the receiving end. Unfortunately, the Hughes system was even more subject to breakdown and synchronization problems than the ABC telegraph, and before long the elegantly simple Morse system won out.

Sometime before the last Hughes telegraph disappeared, American Christopher Latham Sholes visited a telegraph office and observed the Hughes keyboard in action. Inspired by the machine's ability to print characters on paper, Sholes returned to his workshop and developed a similar but entirely different device. On June 23, 1868, Sholes was granted a U.S. patent for the typewriter.

Hughes Telegraph
Signed "Dumoulin - Froment a Paris"
French, Mid 19th Century

Operated on the Paris - Milan line

The Morse Electromagnetic Telegraph

Samuel Morse was first struck with the idea of an electromagnetic telegraph in 1832 while on a return trip from Europe onboard the packet ship Sully. Unfortunately, he made little progress in developing the idea for the next several years. Beset by poverty, he was forced to devote most of his time to painting in order to provide himself with food and shelter. In 1835, Morse became employed as Professor of Literature at the University of New York. This provided him with the security he needed in order to further develop his telegraph.

On May 24, 1844, Samuel F. B. Morse dispatched the first telegraphic message over an experimental line from Washington, D.C. to Baltimore. The message, "What hath God Wrought," had been suggested to Morse by Annie Ellsworth, the young daughter of a friend.

Despite the successful demonstration, the general public remained sceptical. Just as Wheatstone and Cooke had discovered in Europe, a dramatic demonstration was needed to prove the value of the telegraph. That opportunity came shortly at the Democratic national convention in Baltimore. After a controversial struggle, the convention dropped Van Buren, then President, and nominated James K. Polk. Silas Wright was named the Vice-presidential nominee. Wright wasn't present, however, being away in Washington D.C. on business. Hearing of the nomination, Alfred Vail in Baltimore telegraphed the news to Morse in Washington, who passed the message on to Wright. Wright declined the nomination, Morse communicated this back to Vail, and within a few minutes of the nomination, Vail presented Wright's response to the very surprised delegates at the convention. At first they refused to believe that Vail had actually communicated with Wright in Washington, but after confirmation they quickly acknowledged the miraculous power of this amazing new invention.

Working closely with his assistant Alfred Vail, Morse demonstrated his first working model in 1835. Instead of using needles, as Wheatstone and Cooke had done, he devised a simple code and used pulses of current through an electromagnet to deflect a pendulum. A pencil attached to the pendulum made written marks on a strip of paper. Vail soon simplified and improved the code and the apparatus, and a public demonstration was given in 1838. It would be another six years, however, before Congress funded $30,000 to construct an experimental telegraph line from Washington to Baltimore, a distance of 40 miles.

The Morse Pen Register
Source: The Century Magazine, 11/1887 - 4/1888 p. 926

What hath God Wrought?

The first telegraph message sent over the experimental line between Baltimore and Washington D.C., May 24, 1844

The Morse Telegraph System

By the mid 19th century, the Morse telegraph had evolved into a simple and efficient system made up of four basic components: Keys, registers, sounders, and relays.

HOW IT WORKS
As shown below, tapping the key opens and closes an electrical circuit that includes the key itself (at the transmitting or sending end), a battery, and the receiver, which can be either a register or sounder. A relay may be used as well.   When the circuit is opened and closed by the key, pulses of current flow through the circuit. These pluses appear as dots and dashes on the paper tape of the register, or as clicks, if a sounder is used. A trained telegraph operator can translate these marks or clicks into characters of the alphabet, in order to receive a message.

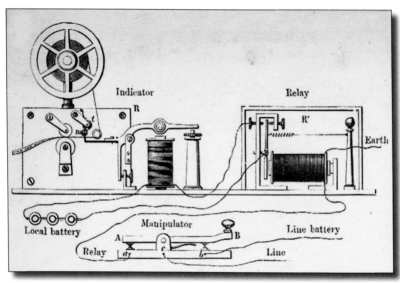

Diagram of a simple Morse telegraph system.   Source: Electricity and Magnetism, Guillemin, 1891

KEY
The telegraph key is the "transmitter" of the system (it is called a manipulator in the above diagram.) See the facing page for detailed information on the evolution of the telegraph key.

REGISTER
The first Morse telegraph receivers recorded the incoming message on a strip of paper tape. Morse felt it was important to have a written record of the original message as received, and didn't dream that operators would eventually learn to "read" the clicking of the register directly. Eventually most registers were replaced by the simpler and much less expensive sounder.

SOUNDER
A sounder consists of two small electromagnets and an iron lever. When the sounder receives a pulse of current, the electromagnet pulls the lever downward, producing a sharp "click".   A spring returns the lever to its original position.   By listening to the sounder, a trained operator can recognize the spacing of the clicks as the dots and dashes of Morse code.

RELAY
A relay is a simple switch operated by an electric current. The design is very similar to that of a sounder, except that a pair of contacts are connected to the lever so to close when the relay is activated. In the diagram above, the relay is used in conjunction with a battery to "isolate" the receiving or "local" circuit from the main telegraph line.   A special form of telegraph relay, called a "repeater," is used to strengthen telegraph signals sent over long distances.

Evolution of the Telegraph Key

The strange looking device at the right is a "port-rule," the first telegraph transmitter developed by Morse and his assistant, Alfred Vail.   It is a good example of how inventors frequently rely on familiar models. This apparatus was built using Morse's knowledge of printing press technology, complete with composing stick and movable type.

Morse Port Rule
Reproduction of original version

Each toothed strip of brass represents a digit, 1-9. Morse developed a code whereby words and phrases were represented by a sequence of numbers. A code book was then used to convert between the codes and the words they represented. For example, the sequence "8732" might mean "reply requested."

Once the "type" is placed on the stick, the message is transmitted by turning the crank. This causes the lever that rides on the top of the type to go up and down, following the pattern of notches on each piece of type. On the left end of the lever are two copper wires that dip in and out of cups filled with mercury, which conducts electricity. The wire dipping into the cup completes an electric circuit, sending pulses of current down the wire to the receiver.

Vail Finger Key
Reproduction of original version

The idea of using a key for transmission didn't occur right away. During tests of the Baltimore-Washington line, Vail began sending code by dipping the wires into the mercury cups by hand - a method he found much easier and faster than arranging pieces of type in the port-rule.   Shortly after he developed the first crude telegraph key, which he called the "finger key."

Not long after he developed the finger key, Vail designed an improved version called the "Lever Correspondent."   This is the key that was   used in the famous 1844 demonstration between Washington D.C. and Baltimore, along with a new system of signals developed by Vail known as the "Morse Alphabet," or "Morse code" as it is more commonly known today.

Vail Lever Correspondent
Replica of 1844 version

European Camelback Key
Mid 19th Century

At left is a "camelback" key from the mid 19th century. Named for their characteristic "hump," camelback keys were among the first commercially used telegraph keys.

ELECTRIC PHRENOLOGY APPARATUS. See page 63
Italian, c. 4th Qtr, 19th Century.

VERY EARLY S.F.B. MORSE DESIGN WEIGHT-DRIVEN TELEGRAPH REGISTER, signed "J.W. Norton, 175 Broadway" Also stamped "153". Dark walnut base with very fancy walnut trim molding measures 12" x 6" and brass open-works frame is 6-3/4" at the highest point. In "Samuel F.B. Morse, His Life and Letters" the frontispiece photograph in Vol. 2 is of S.F.B. Morse holding this exact Norton register in his hand. Original hand made chain is also present on this register. American, c. 1851.

EARLY S.F.B. MORSE DESIGN WEIGHT-DRIVEN TELEGRAPH REGISTER, unsigned. Provenance: from the Charles Came collection of philosophical and telegraph instruments - see Rittenhouse Vol. 5, No. 4 1991 pp. 118-128. Walnut base measures 12 x 5" and is 4³/⁴" in height. Made from cast-iron with original green paint, brass and steel. The U-shaped electromagnet and binding posts, as well as the provenance suggests manufacture date c. 1850.

EARLY S.F.B. MORSE DESIGN WEIGHT-DRIVEN TELEGRAPH REGISTER signed "C. Williams, Jr., Boston." On a walnut base measuring 14 x 7" and is 10" at highest point of the open-frame brass work. Elaborately turned and cut brass, with large gutta-purcha electromagnetic coils. American, c. 1850s.

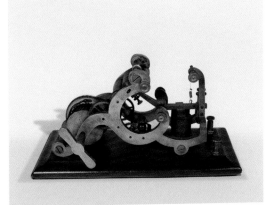

EARLY S.F.B. MORSE DESIGN WEIGHT-DRIVEN TELEGRAPH REGISTER, signed "Tillotson and Co. 16 Broadway, New York." Walnut base measures 13 x 5³/⁴" and brass open-works frame is 7" at highest point. The number "1" is stamped in numerous places on the brass-work. American, c. 1860s.

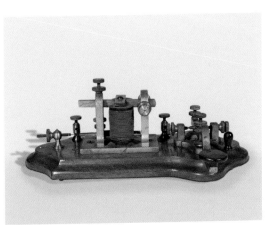

CAMELBACK KEY AND SOUNDER ON BOARD, signed "Cooperative Mfg. Company, Philad, PA". Mahogany base (asymetric contour) measures 11", 6" at longest and widest points. American, c. 1860s.

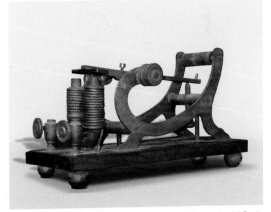

DEMONSTRATION TELEGRAPH REGISTER, unsigned but by Daniel Davis. Illustrated on p. 74, Fig. 33 in "History, Theory and Practice of the Electric Telegraph" by George Prescott, Boston 1860. Rosewood base is 8¹/⁴ x 3³/⁴" and is supported by four brass ball feet (one missing). The object is 5" at its highest point. One foot and the stylus lever are reproductions. American, c. 1840s.

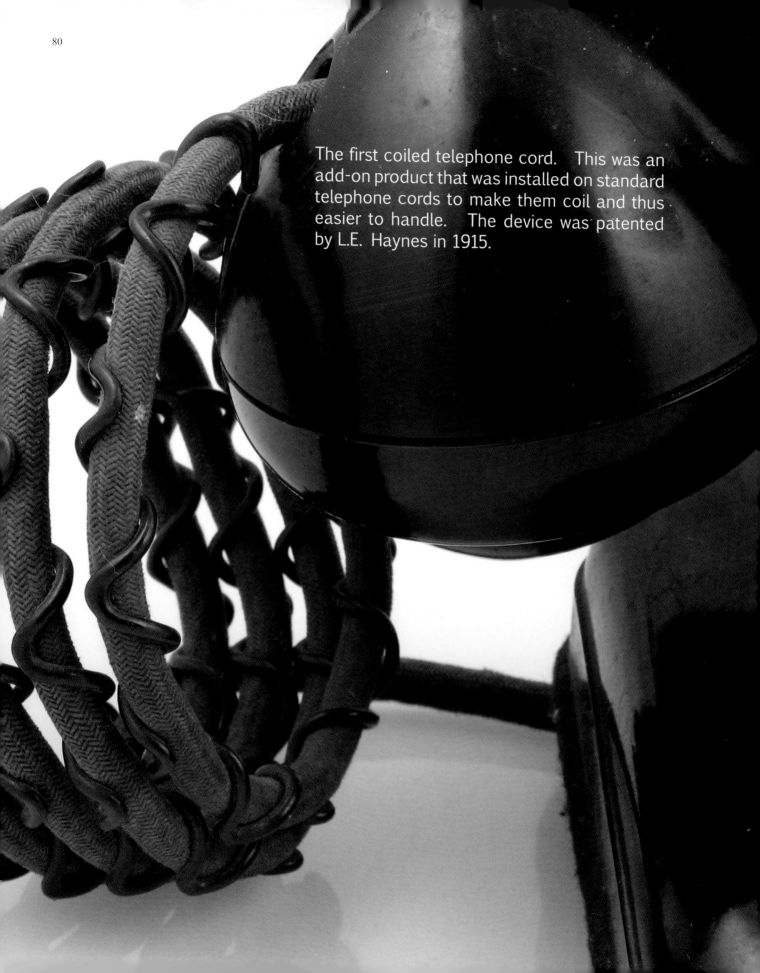

The first coiled telephone cord. This was an add-on product that was installed on standard telephone cords to make them coil and thus easier to handle. The device was patented by L.E. Haynes in 1915.

The Telephone

The telephone was one of the most significant inventions to follow the discovery of electromagnetism. The museum's telephone display includes a representative sample of apparatus dating from before the invention of the telephone and continuing into the 20th century.

Of particular note is the "Higginson" telephone, which was used in a conversation with Bell's assistant Thomas Watson to open the first transcontinental telephone line in 1915 (See page 86.)

Another highlight of the museum's collection is the "Eiffel Tower" telephone (below) made by L.M. Ericsson & Co. Also known as the "Skeletal" telephone, these beautiful desk telephones were among the first to feature a single handset.

"Eiffel Tower" Telephone
L.M. Ericsson & Co.
(Swedish)
c. 1892

BELL "GALLOWS" TELEPHONE.
Bell's first telephone didn't work, but it did secure him a patent and a virtual monopoly on the telephone business for many years. Replica of 1875 Invention.

BELL LIQUID TRANSMITTER.
This is the same type of telephone that Bell was using when he spoke his famous words: "Mr. Watson, come here, I want you." It was Bell's first working telephone, which some scholars argue was invented by Elisha Gray, not Bell. Replica of 1876 model

BELL "CENTENNIAL" TELEPHONE.
Demonstrated by Bell at the 1876 Philadelphia Exposition. Sir William Thomson (later known as Lord Kelvin) observed the demonstration and called the invention " The most wonderful thing in America." Replica of 1876 model

HUGHES MICROPHONE. Unsigned, but likely Pericaud or Radiguet. In 1878, the carbon microphone was invented by David Edward Hughes. Hughes' microphone was the early model for the various carbon microphones now in use. This apparatus was manufactured to be used as an instrument of physics experiments for schools and universities. French, c. 1900

WATTS STRING TELEPHONE. Many people aren't aware that string telephones ("two tin-cans and a string") were a commercially available alternative to the electric telephone. While the sound quality was poor by today's standards, they were a realistic alternative to the expensive and noisy electric telephones of the 1880s. American, c. 1881

STROMBERG CARLSON 2-LINE MAGNETO TELEPHONE.
With "Adjustaphone" scissor attachment made by the Chicago Writing Machine Co. American, c. 1903

Model 288 "Fiddleback" Magneto Wall Telephone
Western Electric Co.
American, c. 1898

Magneto Wall telephone
JYDSK
Danish, c. 1918

Magneto Wall telephone
Kellogg Switchboard and Supply co.
American, c. 1890s

WESTERN ELECTRIC MODEL 102 Round base
The Western Electric model 102 was the Bell system's first handset telephone. This desk set is equipped with the early seamless "spit-cup" E1 handset. American, c. 1924

WESTERN ELECTRIC MODEL 202 Oval base. The Western Electric model 202 was the Bell system's second handset telephone. This desk set is equipped with the early seamless "spit-cup" E1 handset. American, c. 1930

AUTOMATIC ELECTRIC ROUND BASE ROTARY DIAL MONOPHONE. This model, dubbed the "Shirley Temple" phone by collectors, is the first handset telephone used by the independent telephone companies. American, c. 1934

GRAY TELEPHONE PAY STATION MODEL 14.
In 1889, the first public coin telephone was installed by inventor William Gray at a bank in Hartford, Conn. It was a "postpay" machine (coins were deposited after the call was placed.) Gray's previous claim to fame was inventing the inflatable chest protector for baseball. American, c. 1911

STROMBERG CARLSON NON-DIAL CANDLESTICK
This upright desk telephone was Stromberg Carlson's first straight shaft candlestick telephone. American, c. 1908

AUTOMATIC ELECTRIC STRAIGHT-SHAFT DIAL CANDLESTICK. This was the first rotary dial upright desk set used by the independent telephone companies. American, c. 1921

BLAKE TRANSMITTER.
The Blake transmitter was the first successful telephone transmitter (microphone). It was invented by Francis Blake in 1878 and used a carbon and platinum resistance element. It was later improved by Berliner and remained the standard for many years. Base not original. American, c. 1883

STROWGER 11-DIGIT DIAL WALL PHONE. Signed "Automatic Electric." Magneto version of the first dial telephone. See "Strowger 11 digit Potbelly Dial Candlestick". American, c. 1907

STROMBERG CARLSON TAPERED SHAFT OIL-CAN CANDLESTICK. Nickel plated, tapered shaft, aka "oil can" desk set. This rare set is known as the oil can because of the way it looks upside down. It is equipped with the ornate, well marked faceplate and the very hard to find "brass bottomed" receiver. Tapered shaft upright desk sets were the second form of "shaped" candlesticks. American, c. 1900

"girl-less, cuss-less, out-of-order-less, and wait-less."

- Inventor Almon Strowger announcing
the first automated telephone exchange.

STROWGER 11 DIGIT POTBELLY DIAL CANDLESTICK. This is the first dial telephone. The inventor, Almon Strowger, was an undertaker in Kansas City in the late 1800s. According to legend, the wife of his only competitor worked the switchboard at the local telephone exchange. Whenever a caller asked to be put through to Strowger, she would instead deliberately put the call through to her husband, his competitor. After spending years complaining to his local telephone company, Strowger found a way to solve this problem by developing the first automated telephone switch out of electromagnets and hat pins.   When his system made its debut, Almon Strowger bragged that his exchanges were "girl-less, cuss-less, out-of-order-less, and wait-less." American, c. 1905

SOCIETE INDUSTRIELLE DES TELEPHONES
Desk Telephone
French, c. 1919

SOCIETE INDUSTRIELLE DES TELEPHONES
"Trefle" Telephone
French, c. 1890s

The "Hush-A-Phone"

This unassuming little device, called a "Hush-A-Phone," is a small rectangular baffle that fits over the mouthpiece of a candlestick telephone. There is an opening in the front just big enough to place your lips into. When you speak, the party on the other end of the line can hear clearly, but no one in the room with you can make out a single peep.

In the late 1940s the Hush-A-Phone company was sued by AT&T, who didn't allow third-party add-ons to their telephone equipment. The case was eventually decided in 1956 in favor of Hush-A-Phone, a defining moment in the development of an aftermarket for telephone equipment, which eventually lead to telephone modems. Who would have thought this nondescript little black box would be the first step in a long chain of events that would eventually lead to the breakup of AT&T and the development of the public internet?

HUSH-A-PHONE. Signed "Hush-A-Phone Corporation." Shown installed on an Automatic Electric Candlestick Telephone. American, c. 1921

The Higginson Telephone

IT WAS A WESTERN ELECTRIC MODEL 20L – the most common candlestick phone on the market. It looked pretty much like any other candlestick phone - the type used by Sam spade in the *Maltese Falcon* - but this one had a small brass plaque attached to the neck which read:

"This instrument used by Maj.Henry L. Higginson at Boston, Mass. to open the Transcontinental telephone line with Thomas A. Watson at San Francisco, Cal. Monday evening January 25, 1915. Transmitter cutout & signal buttons added"

Intrigued, I asked the owner about it. He said it was sold to him by another collector who had obtained it from a relative of Higginson. I hadn't heard of Higginson, and knew little about the transcontinental telephone line. But this was too good to pass up, so I bought the phone and began my research.

The audience in the AT&T theatre, Panama-Pacific Exposition, San Francisco.

Henry Higginson

Henry Lee Higginson was a noted Bostonian banker and Philanthropist. As a young man during the Civil War, he and his seven friends joined the army, where he served with distinction and attained the rank of Major. Six of the seven friends were killed in the war, a terrible personal loss that

Preparing to send the tones of the Liberty Bell from Philadelphia to San Francisco via the transcontinental telephone line.

would profoundly shape the rest of his life.

An avid music lover, Henry Lee Higginson founded the Boston Symphony Orchestra in 1881 and was its chief benefactor. In 1890, by now a successful banker, Higginson donated 31 acres to Harvard University. He dedicated the gift to his fallen friends, asking that the property be called "Soldiers Field." Today, a large marble marker at the field's entrance recognizes Higginson's gift, and the friends that he held so dear.

Higginson's philanthropy was deeply rooted in the sense of honor he felt for his lost friends. In a letter to Historian James Ford Rhodes, he said: "If my nearest and dearest playmates had lived, they would have tried to help their fellows, and as they have gone before us, the greater need for me to try – and the many tasks are still before us ..."

Higginson's connection to the telephone came through the business of his firm, Lee, Higginson and Company. Because the firm was one of the early financial backers of American Bell (which became American Telephone and Telegraph in 1900), Higginson was invited to participate in the events around the first transcontinental telephone call. The call took place between New York and San Francisco on January 25, 1915.

The Transcontinental Telephone line

The transcontinental telephone line linking the Atlantic seaboard with the West Coast was completed in the summer of 1914. Over 13,600 miles of No. 8 copper wire were laid; four wires crossing 13 states on 130,000 poles. Six repeater stations featuring the new De Forest audion vacuum tube amplifier were required to maintain the signal at acceptable levels. The rate for a three minute call: $20.70

Strict orders were given that AT&T president Theodore Vail's voice must be the first to be heard across the line. This lead to some very creative testing procedures, which ensured no single engineers' voice was carried coast to coast. Finally, on July 29, 1914, with little fanfare, Vail spoke the first words to be heard across the continent. Officials had planned for the launch of the new line to coincide with the opening of the Panama-Pacific Exposition in San Francisco, but since the line was finished a few months early, the public event opening the line had to wait.

The Panama-Pacific Exposition

In 1906 San Francisco was devastated by a great earthquake and fire. Only nine years later, the Panama-Pacific Exposition opened its gates - not so much as a tribute to the completion of the Panama Canal as it was a grand celebration of the rebirth of the city. And grand it was. The 11 exhibit palaces covered over 64 acres. An actual Ford assembly line was set up in the Palace of Transportation and turned out one shiny black Model-T every 10 minutes for three hours every afternoon. The entire area was illuminated by the latest developments in indirect lighting by General Electric. Thomas Edison, Henry Ford and other greats were seen frequenting the grounds of the fair. On opening day, President Woodrow Wilson used a wireless apparatus from his office in Washington D.C. to start the Diesel-driven generator that supplied all of the direct current used in the Palace. There was excitement and wonder in the air.

The magic continued on January 25, 1915, when the 3.400

miles separating New York and San Francisco suddenly vanished as the transcontinental telephone line was officially opened for business. Thomas Watson, Alexander Graham Bell's former assistant, assembled with a group of dignitaries at the Expo's AT&T theatre, while Bell led a similar group in New York. Audience members at both locations were each provided a set of headphones, giving them a firsthand opportunity to listen in.

At 4:30 PM in New York, Dr. Bell lifted the receiver and began a conversation with Thomas Watson.

"Hoy! Hoy! Mr. Watson! Are you there? Do you hear me?"

"Yes, Dr. Bell, I hear you perfectly, do you hear me well?"

"Yes! Your voice is perfectly distinct. "

Later in the call, AT&T President Theodore Vail spoke from Jekyll Island, Ga., and President Wilson offered his thanks and congratulations from the White house. The call continued for some time, with congratulatory speeches and conversations from officials on both coasts. At one point during the call, someone asked Professor Bell if he would repeat the first words he ever said over the telephone. He obliged, picking up the phone and repeating "Mr. Watson, come here, I want you." To which Watson, in San Francisco, replied, "It would take me a week now!"

Henry Lee Higginson and a group of officials waited in Boston. In front of them sat the latest in telephone technology, a Western Electric Model 20AL desk telephone. At 8:00 PM eastern time, Higginson picked up the phone and placed a call to Watson waiting in San Francisco. After exchanging

Front Page of the Boston Globe, January 26, 1915

pleasantries, Higginson handed the phone to Boston Mayor James M. Curley, who spoke with his counterpart, San Francisco Mayor James Rolph. Theodore Vail again joined in from Jekyll Island, and a host of other officials took their turn at participating in this historic event.

The opening events were only a prelude. Exhibitions and demonstrations were staged daily and included remote "conversations" with famous people such as Thomas Edison, Admiral Peary, and many others. An Indian chief spoke from Winnemucca, Nevada, and two Chinese exchanged greetings in their native tongue, offering a simple but effective demonstration that the line could transmit a foreign language. Visitors were also treated to the sound of the surf crashing on the rocks of the Atlantic Ocean. One of the most impressive demonstrations took place in Independence Hall, Philadelphia. A telephone transmitter was placed inside the Liberty Bell, and when it was tapped with wooden mallets, the ring of the old bell was heard in San Francisco. It broke a silence of 80 years, the bell having cracked while tolling the death of Chief Justice Marshall in 1835.

The transcontinental telephone line show at the AT&T theatre would be one of the most popular exhibits of the fair, from opening day until the gates closed on December 4th, 1915. Following the fair, the line continued to capture the public's imagination as heard in The Ziegfeld Follies' "Hello, Frisco," the most popular tune of 1915.

Western Electric 20AL telephones, like the one Higginson used, were introduced in 1915 and made by the millions. But a small brass plaque attached to the neck makes this one unique: A tribute to that magical day when east met west, when the peal of an historic old bell in Philadelphia was heard all the way to San Francisco - the opening of the transcontinental telephone line.

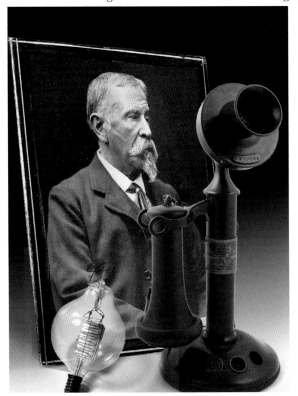

Henry Lee Higginson with his telephone. Also shown is the De Forest Long-Plate audion vacuum tube that helped make the long distance line possible. All are on display at the Museum.

$$\oint \mathbf{E} \cdot d\mathbf{A} = \frac{q_{nc}}{\varepsilon_0}$$

$$\oint \mathbf{B} \cdot d\mathbf{A} = 0$$

$$\oint \mathbf{E} \cdot d\mathbf{s} = -\frac{d\Phi_B}{dt}$$

$$\oint \mathbf{B} \cdot d\mathbf{s} = \mu_0 \varepsilon_0 \frac{d\Phi_E}{dt} + \mu_0 i_{enc}$$

Maxwell's Equations
of the Electromagnetic Field

GALLERY THREE
The Wireless Age

The Wireless Age

Electrical inventions are quickly put to everyday use, delivering lighting and appliances that transform the American household.

Inquisitive inventors break new ground and prove that electrical energy moves through the air like ripples in a pond. Scientists work intently to harness these waves, successfully transmitting and receiving radio signals over long distances.

Wireless radio gets its start in the shipping industry. But after the Titanic disaster of 1912 and WWI prove its value, investors fuel exploration of wireless technology. Hobbyists embrace radio, inspiring amateur inventors and broadcasters to take to the airwaves.

Free from wires, sound travels across the country reaching new audiences, the stage is set for a radio in every home.

The sounds of change fill the air!

GALLERY THREE
THE WIRELESS AGE

TODAY WE ARE ACCUSTOMED to hearing the voices of people in far-off places, such as the moon or atop Mt. Everest. It was not so long ago in human history that communications of this type were unimaginable. Without radio, they would remain impossible, even in this modern age. The first efforts to commercialize radio involved ship-to-shore communications, then an attempt to modernize trans-Atlantic communications by replacing underwater cables. However, radio quickly developed into a medium for providing entertainment and news to the general public.

ELECTROMAGNETIC WAVES

In the first half of the nineteenth century, humans were aware of only the visible portion of the electromagnetic spectrum, namely, that portion that encompassed visible light. That changed when James Clerk Maxwell (1831—1879) discerned that light was a form of electromagnetic radiation. Maxwell was the first to describe a theoretical basis for the propagation of electromagnetic waves. He translated Faraday's theories into mathematical forms, and in 1864 published a treatise containing formulas describing electricity and magnetism. This treatise, titled "On a Dynamical Theory of the Electromagnetic Field," contained the famous "Maxwell's equations." A first edition of Maxwell's famous paper is contained in the Museum's library. These equations implicitly required electromagnetic waves to travel at the speed of light. Maxwell's mathematical formulae predicted that radio waves existed and that they could be created by varying the amplitude of an electrical current in a wire.

Heinrich Hertz (1847—1894) is usually attributed with being the first to send and receive radio waves. Hertz proved the actual existence of Maxwell's theoretical waves in 1887, when he succeeded in creating a current in a wire coil by passing a current through another coil several yards away. Hertz also showed that these waves shared certain properties with visible light. Today, we discuss the frequencies of radio waves in units of "hertz," in honor of Hertz's contributions to the development of radio.

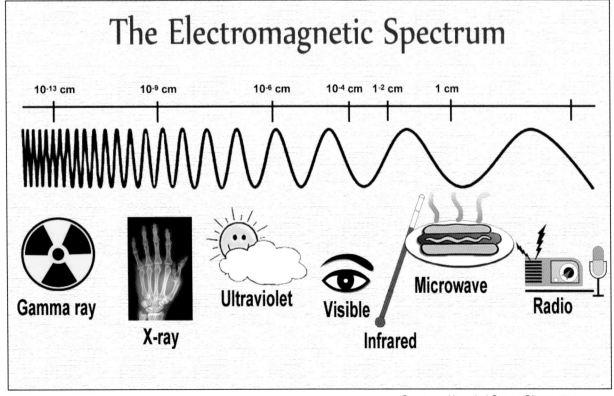

Courtesy: Herschel Space Observatory

EARLY REPRODUCTION OF MARCONI'S 1895 TRANSMITTER. Signed "Comune Di Bologna., No. 38276". Marconi's radio transmitter of 1895 was little more than an induction coil attached to a large metal plate. This reproduction was made sometime in the early 20th century in order to demonstrate his early experiments. Measures 18 x 15³/⁴ x 27³/⁴". Italian, c. 1st Q, 20th century.

WIRELESS TELEGRAPHY

Before radio as we know it today could exist, the scientific observations and theories of Maxwell and Hertz had to be translated into practical uses. An Italian, Guglielmo Marconi

Marconi with his apparatus, shortly after arrival in England

(1874—1937), generally is credited with being the first to develop wireless transmission for practical use. Contenders for the discovery of wireless include Nikola Tesla, an Austro-Hungarian who immigrated to the United States, a Russian inventor, Alexander Popov (1859—1906), and many others.

During Marconi's early youth, long-range communication consisted of telegraph or telephone messages carried by land-lines. Marconi's dream was to bypass land lines by instead using radio waves to achieve point-to-point communication. Marconi's first wireless utilized a simple induction coil that had a fairly limited range. He sought support for further work from the Italian, then the British government, but both turned him down. Marconi next approached the Chief Engineer of the British Post Office, William Preece (1834-1913). Preece himself had tried unsuccessfully to develop a wireless system to communicate with ships, and at once saw the value of Marconi's invention. Preece supported Marconi's work and the two men worked together for several years until disagreements over the best use of radio drove them apart.

Marconi patented his invention in 1896, then formed the Wireless Telegraph and Signal Company Limited. In 1900, this became Marconi's Wireless Telegraph Company, and later, GEC Marconi, a giant multinational company.

Marconi's wireless system utilized Morse Code, the system of dots and dashes that today has been abandoned by all but the most ardent radio enthusiasts. Marconi's company concentrated primarily on ship-to-shore communications, and entirely overlooked the possibility of using radio to broadcast music and the human voice for the purpose of entertainment.

Marconi became determined to prove that he could send messages across the Atlantic Ocean. He knew that radio waves traveled beyond the horizon despite the curvature of the earth, though he did not know why. Later, scientists determined that radio waves do this by bouncing off the bottom of the ionosphere, an electrically charged layer of the atmosphere. Marconi approached the challenge of trans-Atlantic communication by erecting giant radio towers. When windstorms knocked down his equipment on both sides of the Atlantic, he rebuilt them.

Marconi arranged for a test from his high powered spark transmitter at Cornwall, England which he attempted to pick up on a receiver he had constructed in St. John's, Newfoundland. The reported wavelength of the transmission was 820 kHz, and the signal transmitted was a sequence of S's (three dots) in Morse code, sent repeatedly at intervals for a three hour period each day of the testing period. On December 12, 1901, Marconi claimed that he heard the test signal. Accordingly, this date generally is accepted as the first trans-Atlantic transmission, though some radio aficionados have questioned the authenticity of Marconi's claim.

In any case, the signal from Marconi's 1901 transmitter was so weak that it had no commercial value. It took Marconi another five years to obtain a sufficiently strong signal to reliably transmit decipherable messages across the Atlantic. He did this by means of an enormous transmitter that generated a 20,000 volt spark.

The sinking of the Titanic in 1912 brought home the importance of Marconi's invention for the maritime industries. Although the Titanic had been equipped with

Marconi at Signal Hill after successful transatlantic transmission

a Marconi radio and another ship was less than five miles away, it was the Titanic's misfortune that the radio in the nearby ship had been turned off that night. After the luxury liner sank, the use of radio was instantly recognized as being crucial for the safety of ships at sea. New codes were devised and it became standard practice for ships to maintain radio contact at all times.

Marconi eventually returned to his homeland, Italy, where he was received with cries of adulation. Touched by this reception, he opened his affections to his country of origin. He later became a follower of Mussolini, a move that damaged his worldwide reputation. Despite his association with Mussolini, when Marconi died in 1937, his passing was honored by commercial radio stations observing two minutes of radio silence as a tribute to this man's achievements in this field.

The Museum's recreation of the Titanic wireless room

The United States Navy recognized the value of wireless almost as soon as the invention became known. In 1902, the U.S. Navy began to equip its ships with wireless transmitters. Amateurs also began building their own sets, and interference among the numerous transmitters became a constant problem. This was because the spark transmitters used at that time emanated radio waves at a variety of frequencies; the receivers could not "tune out" unwanted signals. When World War I erupted, to avoid interference with Navy communications, the Navy assumed control of all radio transmissions in the United States.

Improvements to radio transmitters came in stages. The first improvement was the arc transmitter, which was a modification of the carbon arc lamp. Carbon arc lamps characteristically emit a humming noise. The arc transmitter was pioneered by an English engineer named William Duddell (1872—1917) who found that by varying the voltage to the lamps, he could control the oscillation of the spark, and thus its frequency. He used his device

to modulate the audible tones emitted by the lamps. However, Duddell considered his invention a curious form of musical entertainment, and did not think his transmitter could oscillate at radio frequencies.

Improvements to Duddell's device were made by Valdemar Poulsen (1869—1942) of Denmark, who in 1902 succeeded in raising the oscillations to normal radio frequencies. He did this by using special electrodes in a sealed chamber filled with a flammable hydrocarbon vapor. At lower energy levels, the frequency could be modulated to allow voice communication.

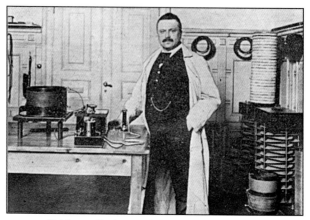

Valdemar Poulsen with his wireless telephone

Thus, the arc transmitter became the first device capable of pure, undamped waves. However, because it employed flammable vapors, the arc transmitter was notorious for occasionally exploding. Moreover, they were difficult to use.

The next improvement in radio transmission was provided by a Canadian, Reginald Fessenden (1866—1932), who sought to compete against Marconi for trans-Atlantic radio communications. Fessenden developed the rotary arc transmitter, which produced a higher frequency than Marconi's transmitters. Fessenden used his transmitter in 1906 for the first transmission of the human voice. Fessenden's device encountered several contacts as it rotated, thus allowing for a continuous production of a spark. Fessenden, together with De Forest (see below), also is attributed with inventing continuous wave amplitude-modulated (AM) radio, which has the advantage of allowing multiple stations to send signals, as opposed to spark-gap transmission, wherein a single transmitter occupies the entire range of radio frequencies.

The Poulsen arc transmitter was

Reginald Fessenden
(1866—1932)

still in use until World War II. The largest arc transmitter was one that operated in Java between 1917 and 1927 and which was rated at 1600-2400 kilowatts. This was far larger than modern transmitters, and arc transmitters produced so much power that their microphones sometimes melted.

VACUUM TUBES

The arc transmitter was eventually replaced by the vastly superior "Audion" vacuum tube, invented in 1906 by American inventor Lee De Forest (1873-1961). The Audion was an improved version of John Fleming's diode vacuum tube detector (see photo on following page.) The improvement was accomplished by placing a third electrode, or "grid," between the cathode and anode of the diode tube; this arrangement allowed the flow of electrons to be controlled by a weak radio signal supplied to the grid. Similar tubes today are known as 'triodes." The Audion could amplify relatively weak electronic signals and also could detect radio signals. The first use of the Audion was in the 1914 transcontinental telephone line. One of these early Audions can be seen in the Museum's telephone exhibit.

THE BOTTLES THAT HOLD THE GENIE
In Dr. De Forest's left hand is the large 250-watt oscillion or "oscillating audion" used for sending radio messages; in his right hand is the small 1-watt audion used for receiving.

By 1916, De Forest had adapted the Audion into the "Oscillion," a tube capable of transmitting radio signals. De Forest did a great deal more for the development of radio, but the Audion tube was his most significant contribution.

The Audion tube was an improvement over the arc transmitter, but it still had its problems. The vacuum inside

the tube was incomplete; atmospheric gases within the tube would inevitably ionize during use to create spectacular colors, followed by failure of the tube. Moreover, the Audion was not capable of linear amplification. De Forest eventually sold his rights to the Audion to American Telephone and

Edwin Armstrong (1890—1954)

Telegraph (AT&T), whose scientists found a way to create a more complete vacuum inside the tube. This was achieved with a "getter," a deposit of barium oxide placed inside the tube which, when heated, would oxidize and leave a silver-colored coating on the inside of the tube. In the process, all the oxygen in the tube was absorbed, thus rendering a much better vacuum. The improved tube was far more stable

De Forest became a prominent, if not notorious, public figure, becoming embroiled in numerous law suits, failed business deals, multiple marriages and even an indictment for fraud. Even so, he was a visionary who predicted the eventual use of microwaves, a form of electromagnetic radiation, as a means for cooking food. However, his vision was less than perfect. He also predicted that the transistor would never replace the Audion amplifier and, believe it or not, that television would never be commercially feasible.

EDWIN ARMSTRONG

De Forest's most extensive legal battle was a patent dispute with Edwin Armstrong (1890—1954) over the invention of the regenerative circuit. A regenerative circuit is one that allows the same signal to be amplified many times by the same vacuum tube. Both men obtained patents that covered this invention. Armstrong had a better understanding than De Forest of how the Audion tube regulated the flow of electricity, and Armstrong expanded on this idea by feeding some of the amplified signal back into the input of the system. This feedback allowed the signal to be further amplified. However, after a protracted battle, the courts

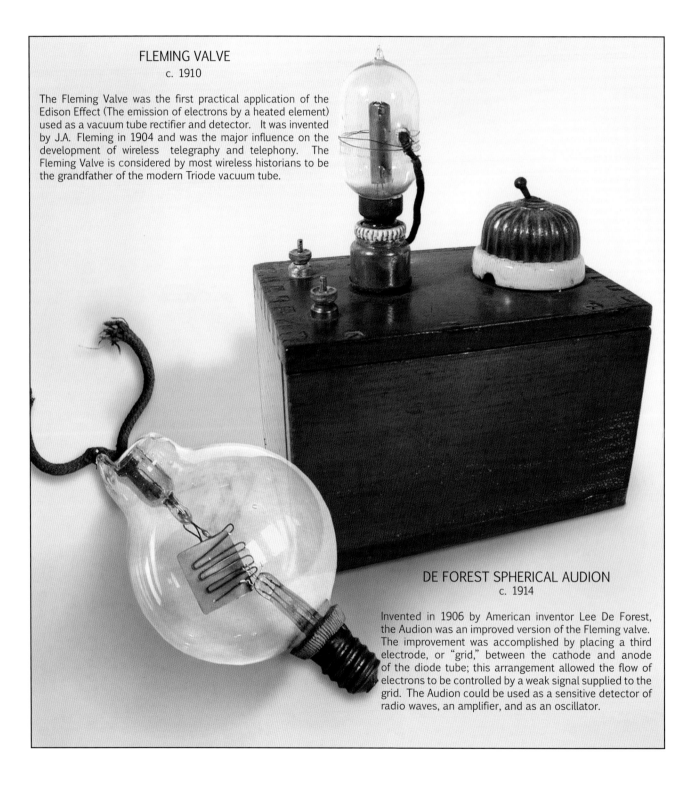

FLEMING VALVE
c. 1910

The Fleming Valve was the first practical application of the Edison Effect (The emission of electrons by a heated element) used as a vacuum tube rectifier and detector. It was invented by J.A. Fleming in 1904 and was the major influence on the development of wireless telegraphy and telephony. The Fleming Valve is considered by most wireless historians to be the grandfather of the modern Triode vacuum tube.

DE FOREST SPHERICAL AUDION
c. 1914

Invented in 1906 by American inventor Lee De Forest, the Audion was an improved version of the Fleming valve. The improvement was accomplished by placing a third electrode, or "grid," between the cathode and anode of the diode tube; this arrangement allowed the flow of electrons to be controlled by a weak signal supplied to the grid. The Audion could be used as a sensitive detector of radio waves, an amplifier, and as an oscillator.

finally found in favor of De Forest's claim to have invented the regenerative circuit, though it is widely believed today that this decision was erroneous.

Armstrong also found that when the amplification was increased past a certain point, the circuit would begin to oscillate and create its own waves. He showed this process to David Sarnoff (1891—1971), an employee of the American Marconi Company. Sarnoff understood the implications of the discovery immediately and urged Marconi to buy the rights to Armstrong's patent. The Marconi company declined. At the end of World War I, the Radio Corporation of America (RCA) was formed, and in 1919, Sarnoff became its commercial manager. Under Sarnoff's direction, the commercialization of radio became fully realized.

Armstrong had invented the "Super Heterodyne" in 1918, while he was in the army and stationed in France. This device combined a low frequency and high frequency wave to form a new wave that could be amplified to much higher levels than before. After the war Armstrong sold his patent to Westinghouse, who cross-licensed to RCA. RCA quickly incorporated Armstrong's superheterodyne design, which provided improved selectivity as compared with De Forest's process. This feature made RCA radios very popular. This, in

David Sarnoff (1891—1971)

turn, made Armstrong very wealthy until his lawsuit with De Forest, and later RCA, ate away his funds.

Armstrong also invented FM radio. Until then radio transmissions were "AM," that is, based on modulating the amplitude of the radio waves. Challenged by Sarnoff to find a way to eliminate the static prevalent with AM radio, Armstrong discovered that he could send transmissions by modulating the frequency of the waves instead of modulating their amplitudes. Frequency modulated, or "FM" transmission, was not subject to outside interference and

produced a much clearer, fuller sound than AM.

Even so, Sarnoff's interests by then had turned to the development of television, and he rebuffed Armstrong's request to invest in FM. Armstrong therefore struck out on his own to set up several FM radio stations in the New England as the Yankee Network. Radios capable of receiving an FM signal reached the market. However, Armstrong's enterprise failed when Sarnoff convinced the U.S. government in 1945 to designate Armstrong's FM frequencies to be reserved for television instead of radio. With one quick move, all the radios that used Armstrong's stations were rendered useless. Armstrong was cut off from the U.S. market, and eventually committed suicide.

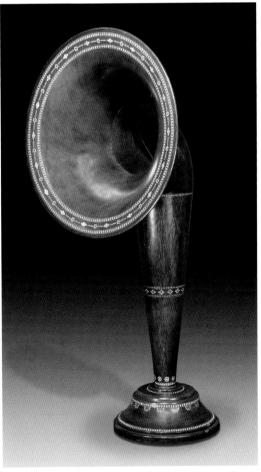

Mahogany horn loudspeaker with inlaid mother-of-pearl. Owned by RCA Chairman David Sarnoff in 1924, now on display at the American Museum of Radio and Electricity.

THE DISCOVERY

MICHAEL FARADAY WAS THE FIRST to propose the concept of an electromagnetic field - lines of force which spread out in all directions to fill space, and to affect matter within that space.

It was James Clerk Maxwell, however, who built the theoretical bridge from electricity to radio. Intrigued by Faraday's work, Maxwell developed a set of equations to describe mathematically how a changing magnetic field produced an electric field (electrical induction). He discovered that the opposite was also true, and that changing magnetic and electric fields in space produced electromagnetic waves. He showed how these waves traveled at the speed of light, and that in fact, light was just another form of electromagnetic wave.

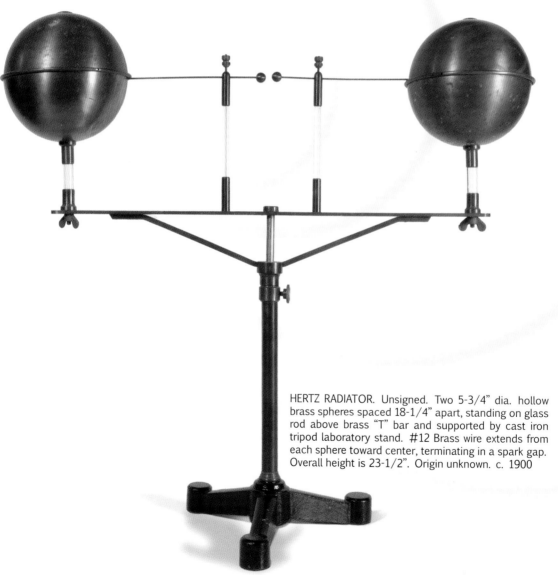

HERTZ RADIATOR. Unsigned. Two 5-3/4" dia. hollow brass spheres spaced 18-1/4" apart, standing on glass rod above brass "T" bar and supported by cast iron tripod laboratory stand. #12 Brass wire extends from each sphere toward center, terminating in a spark gap. Overall height is 23-1/2". Origin unknown. c. 1900

OF RADIO WAVES

Maxwell presented his analysis in 1873, in the two volume books entitled "A Treatise on Electricity and Magnetism."

Little progress toward radio was made until six years later when the Prussian Academy of science offered an award to "establish experimentally any relationship between electromagnetic forces and the dielectric polarization of insulators." A bright young professor of physics at the Engineering College in Karlsruhe, Germany, named Heinrich Hertz picked up the challenge and began a series of experiments.

Hertz used apparatus similar to that shown here. The radiator on the facing page was connected to an induction coil, so that a continuous spark was present between the small brass spheres at the center of the apparatus. The resonator (at right) was placed on a bench nearby. During the course of his work, Hertz noticed that when the induction coil was activated, a small spark appeared in the resonator. Since the resonator had no physical connection to the induction coil or spark gap Hertz knew he was witnessing a new phenomenon.

Hertz thus confirmed Maxwell's theory. Through a series of ingenious experiments he went on to confirm that these new phenomena were indeed waves, they traveled at the speed of light, and that the waves could be reflected, refracted and polarized, the same as light waves. He was also able to calculate their wavelength and frequency.

Hertz's discovery led to a groundswell of interest by many people, including a young Italian named Guglielmo Marconi.

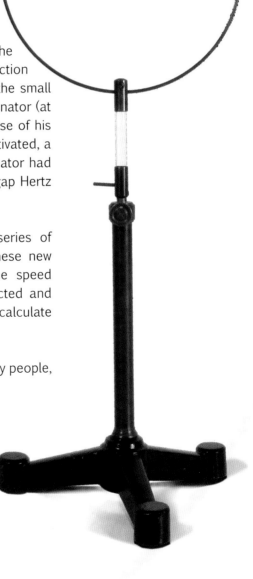

HERTZ RESONATOR. Unsigned. Spark gap formed of 9" dia. brass wire loop terminating in 1/2" dia. solid brass spheres. Loop is held by glass rod and supported by cast iron tripod laboratory stand. Overall height is 23-1/2". Origin unknown. c. 1900

Hertzian Waves

When Maxwell published his famous paper on the nature of the electromagnetic field, the portion of the equations predicting electromagnetic waves was hardly noticed.

It wasn't until more than ten years later when Hertz announced his confirmation of Maxwell's predictions that scientists became interested and an entire new category of apparatus was developed to use in the study of this new phenomena known as "Hertzian waves".

Max Kohl Spark Transmitter (Bose design)
and Coherer Receiver
(German)
c. 1900

Creating Hertzian Waves

To create Hertzian waves, all that is needed is a spark. In fact, the earliest reports of wireless phenomenon or "action at a distance" as it was called, date from the 18th century, when sparks were produced by frictional electric machines or Leyden jars (see pages 12 and 16.) To create sparks you need high voltage, and Hertz used a coil similar to the one below in his early experiments.

WIRELESS RHUMKORFF COIL. Signed "H.W. Sullivan, London." Mahogany base measures 25$^{3/8}$ x 11$^{1/2}$ x 4". 13$^{1/2}$" overall height including spark gap supports. Coil measures 15$^{1/2}$ x 3$^{3/8}$" dia. British, c. 1900

The coil operates like a large electric buzzer: The interrupter's contacts, wired in series between the battery and the coil, are closed in its resting state. When power is applied, the coil is activated, which produces a magnetic field in the iron core of the coil. This pulls the interrupter contacts open, breaking the circuit. The falling magnetic field then relaxes its hold, the contacts close, and the entire cycle repeats.

The rising and falling magnetic field of the primary coil induces a high voltage across the secondary. When the voltage is high enough it jumps the gap, creating a spark. During the time the spark is present there is a high frequency, alternating current electromagnetic field radiated from one pole of the spark gap. This pole is generally connected to some form of antenna.

Detecting Hertzian Waves

Heinrich Hertz used a small spark gap called a resonator (see page 99) in his earliest experiments. Professor Augusto Righi at the University of Bologna developed an improved detector (below). But both detectors suffered from the same drawback; they required direct visual observation and had no mechanism for connecting them to a bell or recorder. That drawback disappeared with the development of the Branley coherer, making wireless telegraphy a possibility.

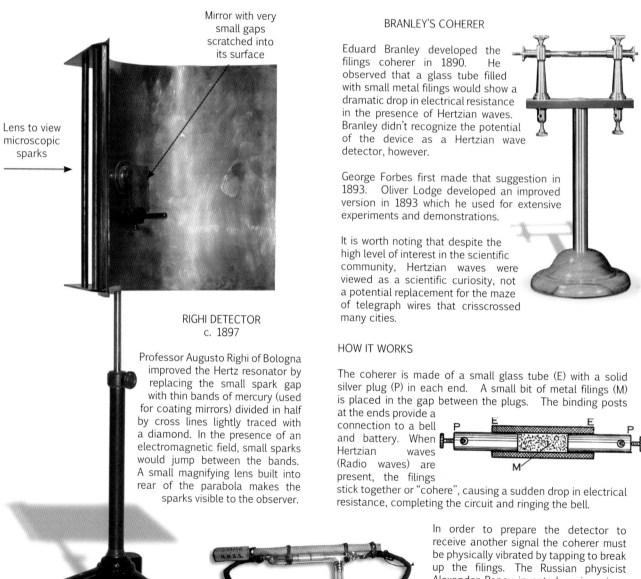

Mirror with very small gaps scratched into its surface

Lens to view microscopic sparks →

RIGHI DETECTOR
c. 1897

Professor Augusto Righi of Bologna improved the Hertz resonator by replacing the small spark gap with thin bands of mercury (used for coating mirrors) divided in half by cross lines lightly traced with a diamond. In the presence of an electromagnetic field, small sparks would jump between the bands. A small magnifying lens built into rear of the parabola makes the sparks visible to the observer.

BRANLEY'S COHERER

Eduard Branley developed the filings coherer in 1890. He observed that a glass tube filled with small metal filings would show a dramatic drop in electrical resistance in the presence of Hertzian waves. Branley didn't recognize the potential of the device as a Hertzian wave detector, however.

George Forbes first made that suggestion in 1893. Oliver Lodge developed an improved version in 1893 which he used for extensive experiments and demonstrations.

It is worth noting that despite the high level of interest in the scientific community, Hertzian waves were viewed as a scientific curiosity, not a potential replacement for the maze of telegraph wires that crisscrossed many cities.

HOW IT WORKS

The coherer is made of a small glass tube (E) with a solid silver plug (P) in each end. A small bit of metal filings (M) is placed in the gap between the plugs. The binding posts at the ends provide a connection to a bell and battery. When Hertzian waves (Radio waves) are present, the filings stick together or "cohere", causing a sudden drop in electrical resistance, completing the circuit and ringing the bell.

In order to prepare the detector to receive another signal the coherer must be physically vibrated by tapping to break up the filings. The Russian physicist Alexander Popov invented an ingenious way to automatically reset the coherer so that it is immediately ready to receive the next signal.

Marconi Coherer
Signed "M.W.T.Co Ltd"
c. 1900

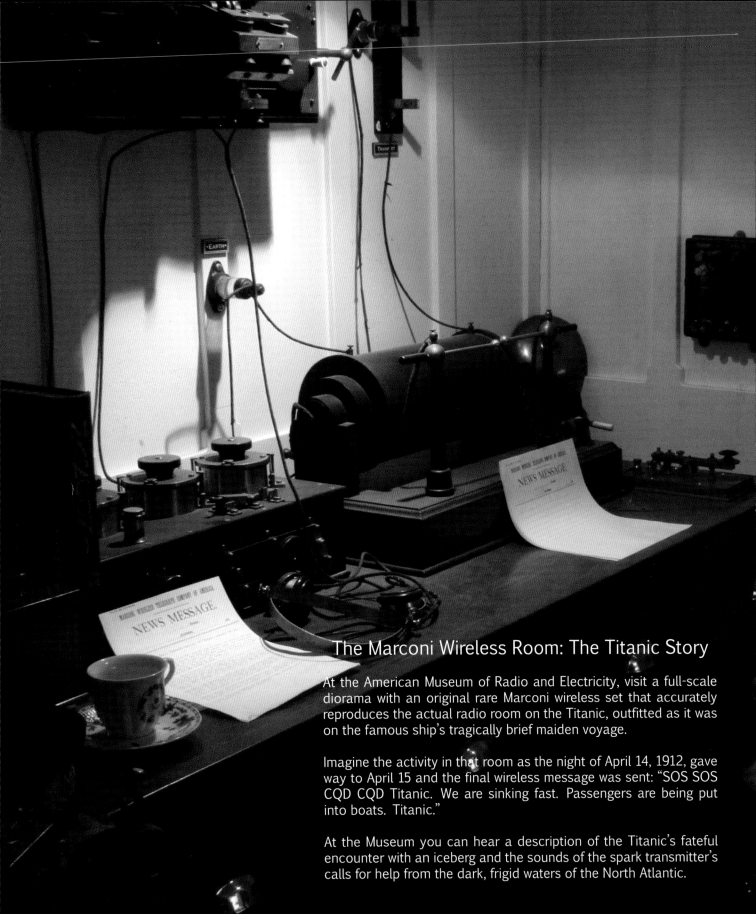

The Marconi Wireless Room: The Titanic Story

At the American Museum of Radio and Electricity, visit a full-scale diorama with an original rare Marconi wireless set that accurately reproduces the actual radio room on the Titanic, outfitted as it was on the famous ship's tragically brief maiden voyage.

Imagine the activity in that room as the night of April 14, 1912, gave way to April 15 and the final wireless message was sent: "SOS SOS CQD CQD Titanic. We are sinking fast. Passengers are being put into boats. Titanic."

At the Museum you can hear a description of the Titanic's fateful encounter with an iceberg and the sounds of the spark transmitter's calls for help from the dark, frigid waters of the North Atlantic.

Hertzian Wave Apparatus

COMPLETE APPARATUS FOR WIRELESS TELEGRAPHY signed "GBN" (Gebruder Bing Ltd). Coherer receiver measures $10^{3/4}$ x $7^{13/16}$ x $6^{1/2}$", Rhumkorff Transmitter measures $6^{1/16}$ x $6^{5/16}$ x $3^{3/8}$". One of the first "toy" radio sets, see 1906 Bing catalogue, p. 318. German, c. 1906.

PARABOLIC HERTZIAN WAVE SET. Unsigned. Parabolic Hertzian Wave Radiator with Righi Spark Gap and Parabolic Hertzian Wave Resonator with coherer detector. Early replica of Marconi's parabolic apparatus. Black enamel wood frame measures $13^{3/4}$ "x $7^{3/4}$ x $20^{3/4}$. Parabola measures 11" wide. Italian, c. 1st Q 20th Century.

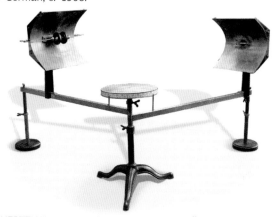

KLICNIK COHERER RECEIVER. Unsigned but by Klicnik Bonn, Germany. Mahogany base measures $10^{3/4}$" x $6^{1/8}$ x $7^{1/2}$".   Coherer with classic decoherer mechanism rings bell when activated. German, c. 1900

RIGHI HERTZIAN-WAVE TEST BENCH, Signed "ARRIGHINI ANGELO, COSTRUTTORE, MILANO." This test bench was designed as a tool to investigate the new phenomena known as "Hertzian waves". Iron tripod at center measures 18" tall, supporting $9^{3/4}$" calibrated walnut disk. Two horizontal brass arms measure $33^{1/2}$", supporting a Righi spark gap on the right side, and a Branley style coherer on the left, each mounted at the focus of a $10^{1/2}$" parabolic antenna The apparatus has many calibrated adjustment points necessary for scientific analysis. Italian, c. 1895.

BALL COHERER. Signed "Epochet D.R.G.M." Mahogany base supported by four wooden feet, measures $3^{5/8}$ x 6". Overall height is $6^{1/2}$".   The Ball coherer consists of a glass tube enclosing six steel balls held in place by a steel spring. The device operates on the "dirty contact" principle discovered by Hughes. An adjustment at the top of the tube provides a way to vary the pressure on the stack, thereby changing the electrical characteristics of the detector. German, c. 4th qtr 19th century

PISTOL FIRED BY HERTZIAN WAVES. Signed "Optiker C. Leibig, Nurnberg", Identical to Max Kohl, see Max Kohl catalog #21, p. 799. Measures $11^{1/2}$ x 6 x $7^{3/8}$". This unusual apparatus is a coherer connected to an electromagnet. When the coherer detects a radio pulse, the magnet pulls the trigger of a cap gun, producing a loud report. This device is believed to be modeled after Marconi's experiments where a gun was used to signal successful reception of a signal transmitted from out of view. German, c. 1905

MORSE PAPER TAPE INKER. Signed "MARCONI'S WIRELESS TELEGRAPH CO. LTD. Mahogany case measures 12 x 10³/⁴ x 6³/⁸". Brass plate on case front indicates no. 58529. Inker is stamped with no. 614.

Early example of one of Marconi's first receiving devices (also known as a telegraph register), The inker receives short and long pulses of current from the wireless receiver, corresponding to the dots and dashes of the received Morse code, and marks a dot or dash on the paper recording tape.

On the base is a clockwork motor, the key of which can be seen. This motor causes the paper tape to pass through the recording mechanism. British, c. 1900.

Marconi Wireless Apparatus

MARCONI MULTIPLE TUNER. Signed "MARCONI'S WIRELESS TELEGRAPH COMPANY Co. Ltd No 26308 LONDON"   Ebonite base supporting mahogany sides, Ebonite front and top measures 19 x 8 x 8$^{1/2}$". British, c. 1910.

MARCONI CRYSTAL DETECTOR/TUNER. Similar to Model 103 but earlier. Signed "MARCONI WIRELESS TELEGRAPH COMPANY OF AMERICA NEW YORK". Ebonite case measures 12"x 12 x 6$^{3/8}$". American, c. 1913.

MARCONI MC1 1.5KW SHIP SET. Signed "MARCONI'S WIRELESS TELEGRAPH COMPANY Co. Ltd No 226619 LONDON TYPE M.C.1." Red panels supported on Aluminum frame. Measures 18 x 12$^{3/4}$ x 30". Right variometer knob, Filament, and choke knobs, and left ammeter needle are reproductions. One of the first Marconi continuous wave transmitters, designed to operate between 2,000 and 3,000 meters. Two rectifier and one oscillation valve. British, c. 1922.

MARCONI MAGNETIC DETECTOR. Signed "MARCONI'S WIRELESS TELEGRAPH COMPANY Co. Ltd No 343 LONDON" with vertical magnets, screw tension adjustment and mahogany base and glass-topped cover - 18$^{3/8}$ x 8 x 10". British, c. 1910.

MARCONI FLEXIBLE DETECTOR/TUNER. Signed "MARCONI'S WIRELESS TELEGRAPH COMPANY Co. Ltd No 49555 LONDON"   Mahogany case, Ebonite top panel. No bottom panel, both leaf-spring crystal detectors damaged, both Billi condenser dielectrics missing. Component of horse-drawn "pack"set. British, c. 1914.

MARCONI TYPE 106D MARINE RECEIVER Unsigned, but by Marconi Wireless Telegraph Company. Black enamel mahogany case, on mahogany base (not original). Ebonite front panel. Original nickel buzzer missing, replaced with newer style.   American, c. 1915.

MARCONI TYPE 106D MARINE RECEIVER. Signed "MARCONI WIRELESS TELEGRAPH COMPANY CO. OF AMERICA".   Mahogany case, Ebonite front panel. Etched brass dial plates.   American, c. 1922.

MARCONI 1/2 KW QUENCHED MULTIPLE SPARK TRANSMITTER. Signed "MARCONI'S WIRELESS TELEGRAPH CO. OF AMERICA TYPE Q.M.S SERIAL No 53" Mahogany case, Ebonite front panel. Measures 14 x 8 x 14". Designed to be used for yachts and small ships in conjunction with the 106B receiver. American, c. 1920.

MARCONI TYPE 16 BALANCED CRYSTAL RECEIVER. Signed "MARCONI WIRELESS TELEGRAPH COMPANY OF AMERICA NO 80087 LONDON". Dual crystal detectors, Ebonite top panel, Mahogany case, Measures 12³/₈ x 10³/₄ x 5¹/₈". This was the standard receiver for 5-KW ship and shore installations. British, c. 1915.

MARCONI 104R "PLAIN TUNER". Signed "MARCONI'S WIRELESS TELEGRAPH COMPANY Co. Ltd No 63877 LONDON" Mahogany base supporting mahogany sides and back, Ebonite front and top. Known as "Plain tuner" due to lack of intermediate circuit. British, c. 1910.

MARCONI TYPE 226A MARINE RECEIVER. Signed "MARCONI'S WIRELESS TELEGRAPH COMPANY Co. Ltd MARINE TUNER TYPE 226A INST No 291790 P.S. No 1754A" Mahogany case supporting ebonite front panel. British, c. 1916.

MARCONI THREE VALVE AMPLIFYING DETECTOR FOR SHIPS. Signed "MARCONI'S WIRELESS TELEGRAPH COMPANY Co. Ltd No 194212 LONDON TYPE No 71" Mahogany case supporting Ebonite top. British, c. 1916.

MARCONI WIRELESS DIRECTION FINDER. Signed "MARCONI'S WIRELESS TELEGRAPH COMPANY Co. Ltd NO 49555 LONDON" Ebonite case supporting radiogoniometer, tuning condenser, et al, measures 17 x 12 x 10¹/₄". British, c. 1910.

MARCONI TYPE 11 DIRECTION FINDER. Signed "MARCONI'S WIRELESS TELEGRAPH COMPANY Co. Ltd LONDON" Teak case with ebonite upper panel supporting seven DEV/Q valves above a sloping panel with radiogoniometer and tuning condenser, measures 18¹/₄" x 11³/₄ x 17³/₄ British, c. 1920.

MARCONI CRYSTAL DETECTOR/TUNER NO. 107. Signed "MARCONI WIRELESS TELEGRAPH COMPANY OF AMERICA NO 5459 NEW YORK". Cerusite or Carborundum detector, Mahogany case, measures 15 x 11³/₈ x 10³/₄". Ebonite top panel, ebonite tapped inductance panel. Restored slide rheostat. British, c. 1915.

MARCONI TRENCH TRANSMITTER. Signed "MARCONI'S WIRELESS TELEGRAPH COMPANY Co. Ltd No 343 LONDON" Paper label in lid reads (partial): "300-600M TRENCH SET W/T 50 WATT WITH CHANGE OVER SWITCH". Mahogany case, Ebonite top panel. Designed for field use in WWI. British, c. 1914.

MARCONI 10" COIL. Signed "MARCONI'S WIRELESS TELEGRAPH COMPANY Co. Ltd LONDON". Mahogany base, ebonite coil support and cover measures 24¹/₄ x 11³/₄ x 13¹/₄". British, c. 1910.

Other Wireless Apparatus

RADIO RECEIVER TYPE A-1 Signed "INDEPENDENT WIRELESS TEL CO INC. NEW YORK" Serial # 98. Range: 200 - 7500 Meters. Mahogany cabinet with ebonite panel. These sets were installed as lifeboat radios aboard the Ocean Liner Leviathan, which had been transferred from duty as a troop carrier during WWI. As S.S. Leviathan, she was the "queen" of the United States' merchant fleet, and operated in the trans-Atlantic trade into the early 1930s. Interestingly, the ship was originally the German built "S.S. Vaterland", seized in April, 1917 and turned over to the U.S. Navy. American, c. 1920.

APPARATUS FOR RECEIVING SPACE SIGNALS. Signed "STONE TEL & TEL CO. BOSTON, MASS USA." Mahogany cabinet measures 16 x 16 x 12$^{1/4}$". Upper left control is marked "Loud" and "Faint" and is mechanically coupled to the variometer mounted in the upper portion of the cabinet. Upper right control adjusts slide inductance. The two lower controls adjust variable capacitors. This is the only known example of this set. American, c. 1907.

Navy Type Receiving-Set

Wireless Telegraph Apparatus

Navy Type Receiving Set

This is one of the finest receiving sets made, and is, we believe, mechanically and electrically perfect, being built with extreme care of the finest materials obtainable.

Each set consists of a navy type tuner, two rotary variable and one fixed condenser, potentiometer, battery switches, testing buzzer, detector and receivers, mounted in a most convenient form of mahogany cabinet.

Source: Central Electric Co. catalog #28, c. 1912 p.984

KPE - CITY OF SEATTLE HARBOR DEPARTMENT RADIO STATION.
Located at Pier #1, Seattle, Washington. A typical wireless station,
c. 1921. Photo courtesy of Pete Petersen.

H.M. WARNER
SEATTLE

ROLLER INDUCTANCE. Unsigned but by Fessenden. Supported on oak base and measures $9^{5/8}$ x 13 x $5^{1/4}$". See "Wireless Telephony", Fessenden, R.A. Proceedings of the American Institute of Electrical Engineers Vol XXVII, No 7, July, 1908, New York. Plate 20, p. 188. An identical roller inductance is seen on the bench at the far right of the photograph. American, c. 1908

ROLLER INDUCTANCE. Unsigned but red National Radio Museum tag reads "Mfg; National Electric Signaling Co. Type 220T. Ebonite base, measures $8^{1/2}$ x $13^{1/4}$ x 6". Wire is loose on rollers, Ebonite support damaged. American, c. 1910

TWO-COIL SYNTONIZER. Unsigned, but by De Forest Radio Telephone & Telegraph Co.. Mahogany case, measures $9^{3/4}$ x 13 x $6^{3/8}$". Missing center sliding contact and portion of top cover. American, c. 1908.

1/2KW SPARK GAP TRANSMITTER AND RECEIVER. Signed Amalgamated Wireless Australasia (AWA)
Spark Transmitters were widely used in ships during the early 1900s. This set was installed on the S.S. 'Burwah', a merchant ship of Australian registry. The photo on the right shows the radio room of the S.S. Burwah, with the set installed. AWA was formed in 1913 and is still in existence today. Australian, c. 1919

RADIO RECEIVER TYPE IP-501A. Signed "Wireless Specialty Apparatus Co." This receiver was designed for the reception of radio telegraphic signals over the wave-length of 250 to 8000 meters (1200 to 37.5 Kc.). This band may be extended to include 18,000 meters (16.7 Kc.) by the addition of a type IP-503 loading unit. It was used mainly aboard ship and at shore stations from about 1925 through 1940.
American, c. 1920

RADIO RECEIVER TYPE IP-76. Signed "Wireless Specialty Apparatus Co." A fine example of one of only a few of these early receivers that still survive. Oak cabinet measures 22-3/8 x 8-1/8 x 10". American, c. 1910

STANDARD SHIPBOARD RECEIVER, TYPE SE-143. Unsigned, but by National Electric Signaling Co. Range: 100-1200 KHz Oak case measures 22$^{7/16}$ x 14$^{1/2}$ x 11". American, c. 1917.

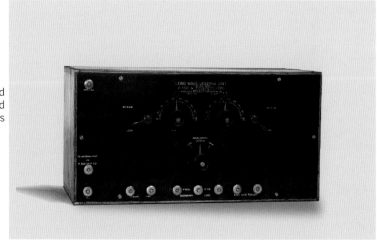

LONG-WAVE LOADING UNIT TYPE IP-503. Signed "Wireless Specialty Apparatus Co." This unit was used in conjunction with the IP-501A to expand the receiver's range to 18,000 meters. American, c. 1920

DE FOUCAULT MERCURY INTERRUPTER. Unsigned but likely by Ducretet . Mahogany base measures 14 x 12". Interrupters such as this were used to provide an interrupted current to the primary of large Rhumkorff induction coils. French, c. 1880

MODEL D TUNER . Signed "United Wireless Telegraph Co." Mahogany case, measures $11^{3/4}$ x $12^{1/4}$ x $5^{1/2}$". Crystal detector is reproduction, otherwise fine condition. American, c. 1910.

MODEL E TUNER . Signed "United Wireless Telegraph Co." The United Wireless Type E tuner was shipped for only a year before Marconi acquired the company in July of 1912. Ebonite panel on Mahogany base measures $12^{1/4}$ x $12^{1/4}$ x $8^{1/2}$". Note missing detector and chip in front corner of ebonite panel.   American, c. 1911.

FESSENDEN WIRELESS SYSTEM WAVEMETER. Signed "National Electric Signaling Co., Pittsburgh, PA USA" Mahogany case, measures $11^{3/4}$ x 12 x $8^{3/4}$". Missing lid, otherwise fine condition. American, c. 1910.

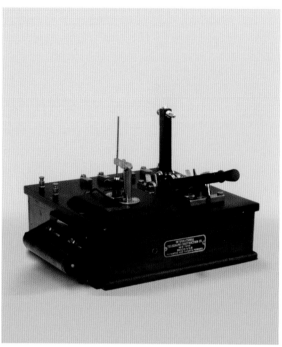

RECEIVING SET - DETECTOR . Signed "Independent Telegraph Construction Co. New York" Mahogany case, measures $13^{3/4}$ x 11 x 7". Electrolytic detector utilizing platinum wire in 20% sulfuric acid solution. Paper tag reads: "SN 1702 from USS Ozark". American, c. 1906.

RECEIVING SET - TUNING INDUCTANCE . Signed "Independent Telegraph Construction Co. New York." Mahogany case, measures $16^{3/4}$ x $9^{1/4}$ x 9". American, c. 1906.

Photo of an Independent Telegraph Construction Co. receiving set, (also known as a Shoemaker set, after its founder) similar to the photos immediately above, showing a similar detector and identical tuned inductance. From Robison's "Manual of Wireless Telegraphy for Naval Electricians", 1906

The Collins Wireless Telephone

A. Frederick Collins
1869 - 1952

It had been two years since Marconi's successful wireless telegraph transmission across the Atlantic ocean, and another year would pass before the invention of the vacuum tube. Wireless Telegraphy, though still in it's infancy, held great promise for the future. Men with names like De Forest, Edison, Fessenden, Marconi, and Tesla were working intensely to make wireless a commercially viable alternative to the wired telegraph.

As the 20th century dawned, so dawned an era of stock speculation, fueled by the millions made by investors in the telegraph, telephone, electric light, and other amazing inventions of the late 19th century. Everyone knew that a wireless telephone would be next, and no one wanted to miss it. For a stock speculator, it was fruit ripe for the picking.

A. Frederick Collins invented his Wireless Telephone in 1898, and formed the Collins Wireless Telephone Company in 1903. He spent the next four years attempting to perfect his device, while at the same time looking for financial backing for his company. To dramatize the potential of this new medium of communication, he began to give public demonstrations of his wireless telephone for businessmen, the press, and the public. In order to support himself and his family, Collins began writing for the Electrical Works, Scientific American, and other technical publications, and delivered lectures for the New York Board of Education on wireless and color photography.

In 1907 Collins met Wall Street promoter Cameron Spear, who had been recommended as someone who could help Collins get his business moving. Under the guidance of Spear and Collins' own attorneys, a public offering of stock was made and public demonstrations were held, often with a celebrity or famous government official participating (see below right.) In December, 1909 Collins Wireless Telephone Company became a part of the Continental Wireless Telegraph & Telephone Company, and at the insistence of his lawyer, A. Frederick Collins was appointed Technical Director. Unfortunately, in the process, Collins lost control of his wireless patents.

The hyperbole surrounding the company was greater than its actual value, and while Collins focused on the technical side of the business, he was apparently unaware that Spear and others were engaging in less than ethical business practices.

THE SEATTLE SUNDAY TIMES
SEPT. 5, 1909.

FOR AUTOMOBILES

The Collins Wireless 'Phone Will Eliminate Many of the Troubles Experienced While Motoring at a Distance from a Garage.

MESSAGE FROM AN AUTOMOBILE.

"Monkey Trial" lawyer William Jennings Bryan and Governor Goldwater participate in a demonstration of the Collins Wireless Telephone. Prescott, Arizona, Sep 19, 1909.

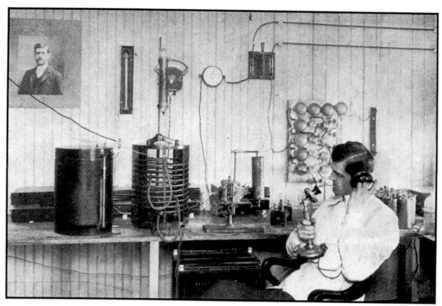

A. Frederick Collins at work in his laboratory

In 1909 the company was raided by the U.S. Post Office, and on January 11, 1911, Collins, Spear, and two other company officials were indicted for using the mails to defraud in selling worthless stock. Collins was charged with giving a fraudulent demonstration of his wireless telephone on Oct. 14, 1909 at the Electrical Show in Madison Square Garden, New York, for the purpose of selling stock in the Collins Wireless Telephone Co.

Collins, Spear, and one other officer were convicted and sentenced on January 10. 1913 to prison terms of up to four years. Collins was freed on parole after one year. Spear served two years before receiving a pardon by President Wilson due to poor health.

Collins claimed innocence at the trial and continued to do so until the day he died. Despite numerous demonstrations of his rotary arc wireless telephone, and testimony by experts that it did, indeed work, the judge remained unconvinced.

Collins rotating oscillation arc prototype

An examination of press coverage and court documents reveals the court's marked lack of technical understanding related to wireless. At one point in the trial, evidence was presented that Collins' "so called" wireless telephone wasn't wireless at all. The evidence? The use of ground and antenna wires.

Before his conviction he had been a respected engineer, considered an authority on wireless in general and a specialist in wireless telephony. And although he was close to perfecting his wireless telephone before his arrest, technology moves at a fast pace. By the time he was able to resume work, Continental Wireless was out of business, and advances such as De Forest's vacuum tube left him standing at the sideline.

After a difficult divorce, Collins left the country and spent several years traveling. In the years that followed, he continued as a prolific writer. His "Radio Amateurs Handbook", first published in 1922 and a staple of every radio hobbyist since, is still in print, now in its 85th edition.

On January 3, 1952, Archie Frederick Collins passed away in his hospital bed in Nyack, New York. He was 83.

As the author of over fifty books, many of them written for a juvenile audience, he touched thousands and thousands of young minds. Who knows how many received the spark of inspiration through his words, how many went on to help create the miraculous world we live in?

COLLINS WIRELESS TELEPHONE (INDUCTIVE MODEL). Signed "Collins Wireless Telephone Co. Newark, New Jersey." Mahogany cabinet measuring 11-7/8 x 8-1/4 x 12" sits 27-3/4" above the floor on brass pole supported by ornate cast iron foot. Microphone extends from the top of the cabinet 15-1/4". A small induction coil and four dry cells are contained inside. Other components including a de-coherer, telegraph relay, and switch are mounted inside top lid. American, c. 1909

ENLIST IN THE SIGNAL CORPS

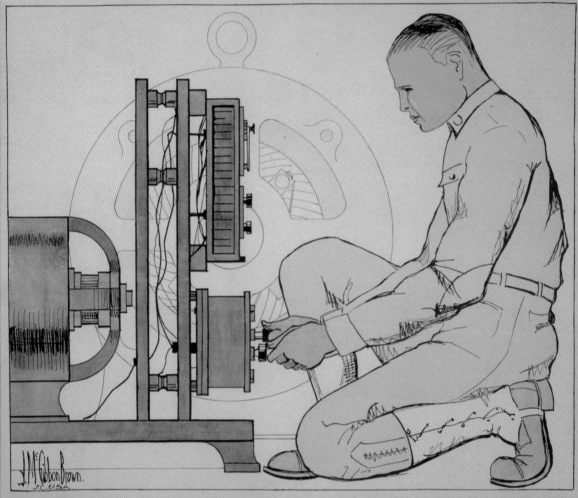

RECEIVE A TECHNICAL EDUCATION FREE

ELECTRICITY - - - -   - - - - - - WIRELESS
TELEPHONY - - - -   - - - TELEGRAPHY
PHOTOGRAPHY - - -   - CARRIER PIGEONS
LINE WORK - - - -   SUBMARINE CABLES
CABLE SPLICING - -   - MOTORCYCLISTS
TEL. CONSTRUCTION   - CHAUFFEURS, ETC.

APPLY TO

GOVERNMENT PRINTING OFFICE

RADIO GOES TO WAR

Before the first world war, military communication had been dependent on runners, flags, carrier pigeons, smoke signals, or at best, wired telegraphy. Radio provided WWI's commanders with quicker, more reliable communication with the soldier in the field.

The radio apparatus seen here represents the "state of the art" c. 1918.

CW-938A SHORT RANGE TRANSMITTER/RECEIVER. Signed "Western Electric Co." The box-jointed, walnut cabinet and ebonite front panel measures 26⁷/₁₆ x 7¹/⁴ x 20". Total weight is 60 lbs. Brass nameplates, corner protectors, dials and switches, are all gold-plated. The left-hand panel section can be removed to expose the transmitter for adjustment purposes. All the internal parts are attached to the back of the front panel, which, when unlatched, pulls down for service. The unit contains five vacuum tubes, including two Type VT-2 and three Type VT-1 tubes. The transmitter is a self-excited oscillator type. The receiver is likely regenerative.. The set tunes from 256 to 600 meters. American, c. 1917.

SCR 59 RECEIVER. Signed "Western Electric Co." Another very early U.S. Signal Corps receiver, designed for use in WWI aircraft. American, c. 1917.

T-1 MICROPHONE. Signed "Western Electric Co." A simple, single-button carbon microphone that was held to the pilot's or observer's chest with a pair of straps to allow hands-free operation. Used with U.S. Army Signal Corps SCR-57, 59 and 68. American, c. 1917.

SCR-68 AIRPLANE RADIO TRANSMITTING AND RECEIVING SET. Signed "Western Electric Co." Measures 17 x 11 x 7". The transmitter section utilizes two type VT-2 tubes as oscillator and modulator. Supports two single-button carbon microphones. Three windows provide viewing the filaments of the three VT-1 tubes in the receiver section. The receiver consists of a non-regenerative detector and two stages of audio amplification. See also SCR-73 (Westinghouse GN4), Western Electric T-1 microphone, and L.S. Brach Antenna Reel. American, c. 1918.

L.S. BRACH ANTENNA REEL. Used in conjunction with the WWI aircraft wireless equipment on this page. The antenna was wound up on the reel when not in use. American, c. 1918.

MILITARY RADIO, Cont'd

CN-112 SHORT WAVE RECEIVER. Unsigned but by National Electric Supply Company (NESCO). This instrument is designed primarily for the reception of spark signals using a crystal detector, or for the reception of short, sustained waves using an audion control box as a detector A four pole, double throw switch in the receiver permits of the use of either audion or crystal detector, both of which may be permanently connected to the receiver. Rage: 300 to 2500 meters. Measures 13 x 7 x $6^{1/4}$". American, c. 1917.

GN4 AVIONICS ALTERNATOR & TRANSMITTER, SIGNAL CORPS TYPE SCR 73. Signed "Westinghouse Electric & Mfg. Co.". This unit was mounted under the wing of the aircraft so that air motion rotated the propeller. Measures 20" long x 6" Diameter at shaft. Propeller (reproduction) measures 28". Note remains of red National Radio Museum tag. See photo of dismantled unit below. American, c. 1918.

Partially dismantled GN4 damped wave airplane radio telegraph transmitting set with synchronous rotary spark gap. From left to right, the self-excited alternator, spark gap, power transformer and oscillation transformer are integrally mounted and enclosed by the micarta streamline casing seen at the right. Extra rotary electrodes for the spark gap for changing the spark frequency are seen at the extreme left. Signal Corps set type SCR73.

SCR 70 ONE-TUBE REGENERATIVE RECEIVER. Unsigned, but by Westinghouse. U.S. Signal Corps communication receiver designed by Frank Conrad. Conrad is considered by many to be the father of commercial radio broadcasting. Range: 0.20-0.50 Megahertz American, c. 1917.

U.S. ARMY SIGNAL CORPS J45 TELEGRAPH KEY. Signal Corps J37 key mounted on a spring steel leg clamp for use in aircraft and tanks. The key is hinged for use on a desk as shown, or it may be flipped over for use when clamped to the operator's leg. American, c. 1945.

MARK III SHORT WAVE RECEIVER by Automatic Telephone Manufacturing Co. Ltd, London. A short-wave crystal receiver used by R.F.C. ground stations during WWI. Mahogany case, outside covered with black painted canvas. Measures 8 x 14 x 12". British, c. 1916.

MILITARY RADIO, Cont'd

BRITISH FIELD RADIO SET (TRANSMITTER). Signed "MARCONI'S WIRELESS TELEGRAPH COMPANY Co. Ltd No 343 LONDON" Paper label in lid reads (partial): "300-600M TRENCH SET W/T 50 WATT WITH CHANGE OVER SWITCH". Mahogany case, Ebonite top panel. Designed for field use in WWI, these sets were too large and heavy to be used at the front lines and the long aerial made the set (and its operator) an easy target. They were replaced for that purpose by the "Forward" sets (see below.) British, c. 1914.

U.S. ARMY BC-14A FIELD RECEIVER.
The BC-14A was copied from the French A-1 receiver by the U.S. Army. It was manufactured by at least four companies: De Forest, Liberty Electric, Wireless Specialty and General Radio. The set was designed for use as an artillery spotting receiver during WWI. American, c. 1917

U.S. ARMY BC-15A AIRPLANE RADIO TRANSMITTING SET.
Signed "Connecticut Telephone & Electric Co."
Aircraft Transmitter, quenched-gap, 1-3 MHz . First transmitter designed for aircraft, Used in WWI. American, c. 1917

W/T FORWARD 20 WATT SPARK TRANSMITTER. These small and extremely portable sets were designed to be carried by the troops at the front lines, to be used primarily for targeting artillery. The compact package and built-in loop antenna made the set ideal for this purpose. British, c. 1917.

The First Radio Designed for Home Entertainment

Westinghouse RA tuner and DA detector/amplifier, Type LV "Vocarola" loudspeaker with "Operola" reproducer. The RADA was Introduced by Westinghouse in 1920 to coincide with the launch of KDKA, the first commercial broadcast radio station.

GALLERY FOUR
Radio Enters the Home

Radio Enters the Home

The magic of early radio broadcasts captivates the country. News is heard on the very day it happens, and opera performed in New York City is experienced live in small towns from coast to coast.

This new medium sparks the creative spirit and attracts entertainers, craftspeople and entrepreneurs.

Shortly after the first broadcasts in 1920, the popularity of radio explodes. Americans eagerly embrace the invisible world of sound that ripples across the land. By decade's end, over 60% of households own a radio.

Radio roars into the 20s!

GALLERY FOUR
RADIO ENTERS THE HOME

THE FIRST RECORD OF PUBLIC BROADCASTING tells of a man named Charles David Herrold (1875 – 1948) who began broadcasting music and other entertainment in San Jose in 1906. Herrold made a distinction between "narrowcasting," which was a transmission destined for a single receiver, such as an offshore ship, and "broadcasting," which meant transmissions targeted at a general audience. He designed omni-directional antennae to help spread his transmissions in all directions. Herrold's transmissions used spark gap technology, but with the carrier frequency modulated to carry the human voice.

The first licensed broadcast radio station, KDKA, was launched in Pittsburgh in 1920. The creation of KDKA was a serendipitous offshoot of a Westinghouse project headed by a researcher named Frank Conrad. Conrad had decided to conduct research testing by placing a receiver at his home while his laboratory assistants sent him signals from a transmitter located at the Westinghouse facility.

Comedian Will Rogers broadcasting from KDKA in Pittsburgh

For these purposes, Conrad arranged for his assistants to play recorded music into the transmitter. Conrad's neighbors, using their personal crystal sets, started listening in on this music, and even began making requests for particular musical selections to be played. Conrad's boss at Westinghouse took note of Conrad's unofficial enterprise and instructed Conrad to build a larger transmitter. As the number of listeners grew, the owner of the local hardware store asked Conrad to announce over the air that crystal radio sets could be purchased at his hardware store. Hence, the first radio advertisement was born.

Early crystal receivers required no power source other than the radio signal itself. However, the signal received by a crystal radio is not amplified and the sound can be heard only through a set of earphones. Moreover, a crystal receiver works only for AM signals and are tricky to tune. Crystal sets were an insufficient basis for building a commercial broadcasting system of the type that exists today.

By the 1920s, radio receivers began to use the same type of vacuum tubes used in broadcasting, which provided many advantages over crystal sets. Vacuum tubes act as amplifiers, picking up relatively weak signals and boosting the output audio signal enough to drive a loudspeaker. Headphones were no longer needed, which meant that more than one person could listen to the same radio. Regenerative circuits, invented in 1914 by Edwin Armstrong, provided a far greater selectivity and sensitivity than could be obtained with crystal radio receivers.

Until the mid 1920s, home radio receivers were not user-friendly. Typically a radio used three different types of batteries employing three separate voltages, known as A, B and C batteries. These batteries needed frequent changing, a dirty and messy job. Tuning in to a station could be quite tedious; Regenerative radios (or "regens" as they were called) were temperamental and to be adjusted constantly during use to avoid oscillations or squeals. When adjusted improperly, the sets could become small radio stations, transmitting annoying squeals and howls into neighbor's radios.

Air Concert "Picked Up" By Radio Here

Victrola music, played into the air over a wireless telephone, was "picked up" by listeners on the wireless receiving station which was recently installed here for patrons interested in wireless experiments. The concert was heard Thursday night about 10 o'clock, and continued 20 minutes. Two orchestra numbers, a soprano solo—which rang particularly high and clear through the air—and a juvenile "talking piece" constituted the program.

The music was from a Victrola pulled up close to the transmitter of a wireless telephone in the home of Frank Conrad, Penn and Peebles avenues, Wilkinsburg. Mr. Conrad is a wireless enthusiast and "puts on" the wireless concerts periodically for the entertainment of the many people in this district who have wireless sets.

Amateur Wireless Sets, made by the maker of the Set which is in operation in our store, are on sale here $10.00 up.

—West Basement

Advertisement from the Pittsburgh Sun
September 29, 1920

By 1923, the most popular radio was the new "TRF" or "three dialer" design. Although they were less temperamental than their predecessor, TRF radios remained battery operated, and they required three knobs for tuning.

The combination of the difficulty in using these radios and the lack of broadcasting content that appealed to a broad audience relegated radios of the early 1920s to the garage or child's room. Because their customers consisted mostly of hobbyists, manufacturers of these early radios paid scant attention to aesthetics, packaging their equipment in ugly utilitarian boxes.

Young boy and friend listening to a crystal radio
c. 1922

With more radio stations crowding the dial, the need for improved selectivity became paramount. By the late 1920s the cumbersome TRF format was pushed out of the market by "Superheterodyne" radio receivers based on an improved technology invented by Edwin Armstrong. Superheterodyne radios could be tuned with only a single dial instead of three, and the invention of AC vacuum tubes meant they could be simply "plugged in." Their improved sensitivity, selectivity, and ease of use finally made radio accessible to everyone, not just the dedicated specialist. RCA quickly began incorporating the Superheterodyne into their units, and RCA radios thus became very popular. Thereafter, the production of radios for the home exploded, as did the broadcasting industry that provided the content for home radio.

The combination of ease of use, radio broadcasts with wide appeal such as news, comedy, and dramatic programming, and the rapidly growing number of stations moved the radio from the garage to the living room and drove a major movement in radio design. By 1930, radios were sold as furniture, and radio specialty shops began to close as the majority of radios were purchased from furniture stores.

A family listens to their new radio, c. 1922

ZENITH 'SUPER ZENITH' HIGHBOY RECEIVING SET, Model VIII   Signed "Zenith Radio Corporation, Chicago, Ill. U.S.A." Early highboy console, 5 knobs, including two tuning controls for volume control and amplification. Features rheostat for filament control and fine tuning switch for additional signal clarity. Original instruction booklet reads 'The exceptional selectivity permits the listener to penetrate the most powerful local broadcasting stations and freely reach out to a vast number of distant stations." Battery compartments on left and right with scrolled metal design, mahogany legs, 6 tubes, battery. 1925 Zenith advertisement reads "It Costs more – because it Does more!" American, c. 1924.

ATWATER KENT CONSOLE, Model 36. Signed "Atwater Kent Mfg. Co. Philadelphia, USA" Wooden cabinet designed and built by Pooley Company with double doors, featuring left front dial, center AK logo, 3 control knobs, 4 ganged tuned RF stages with binocular coils at each stage. Speakers are mounted in top of the cabinet, center front covered with decorative wood grill and matching cloth. 12 inch electro-dynamic speaker, 7 tubes, stands on 4 carved legs. American, c. 1927.

ATWATER KENT HIGHBOY CONSOLE, Model 37. Signed "Atwater Kent Mfg. Co. Philadelphia, USA" Highboy console featuring wood cabinet designed and produced by the Pooley Company. 2 knobs, left front dial, 7 tubes, AC. The Model 37 was At-water Kent's first set completely designed for AC current. This chassis was originally housed in a metal box. As radios moved out of the garage and into the parlor, fine furniture manufactures like Pooley were commissioned to create the cabinets that house these otherwise less attractive chassis. American, c. 1927.

RCA RADIOLA SUPER – VIII CONSOLE, Model AR – 810. Signed "Radio Corporation of America." Popular mahogany lowboy cabinet. Fold-down front revels inner 2 dial panel with antenna controls for tuning built-in multi-directional antenna. The upper cabinet features an internal loudspeaker, which was unusual for the time, and covered with wood cutouts in candlestick design and gold grille cloth. Built in battery storage in lower compartment. 6 tubes. RCA's first console radio and the first of the Radiola Superheterodyne receivers. This popular model sold almost 150,000 its first year. BC, battery. American, c. 1923.

Scroll Radio Receiver
Unknown maker
French, c. 1925

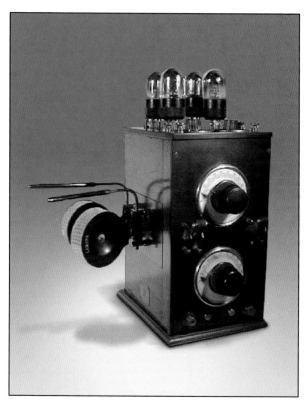

A. Hardy 4-valve Receiver
French, c. 1923

Unidentified Two-Valve Receiver
British, c. 1925

De Forest Model W-5 Receiver
American, c. 1926

These very rare dolls were created by well-known doll artist Joseph L. Kallus, based on a design by the famous illustrator Maxfield Parrish. They were used as point-of-sale pieces to promote RCA and GE radio products. The hands are slotted to hold advertising cards.

RCA ADVERTISING DOLL. "The Sellin' Fool" RCA Radiotron advertising doll. Signed "Cameo Doll Co." Composition head and chest with wooden appendages. 15$^{1/2}$" tall. American, c. 1926.

ADVERTISING DOLL. General Electric Radio Man advertising doll. Signed "Cameo Doll Co." Composition and wood majorette figure measures 19$^{1/4}$" tall. American, c. 1930s.

Crosley "Pup" Receiver with "Bonzo" Mascot
American, c. 1925

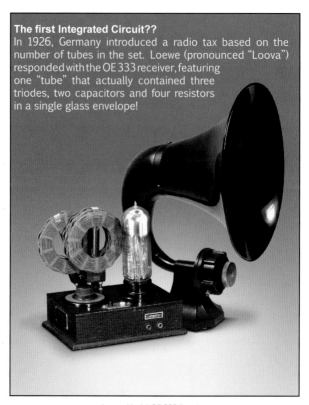

The first Integrated Circuit??
In 1926, Germany introduced a radio tax based on the number of tubes in the set. Loewe (pronounced "Loova") responded with the OE 333 receiver, featuring one "tube" that actually contained three triodes, two capacitors and four resistors in a single glass envelope!

Loewe Model OE 333 Receiver
with loudspeaker
German, c. 1926

Radio Téléphonie Française
3-valve TRF receiver, plate glass panels.
French, c. 1924

RIC
with Grammont Radiofoto type VO valves
French, c. 1923.

Magnavox AC-3 Power Amplifier
with Microphone and R2 Loudspeaker.
American, c. 1923

"PIANO" 4-VALVE RECEIVER. Signed "Société des Établts Ducretet, 75, Rue Claude Bernard, Paris."
Shown with Grammont Radiofoto type 209 Marine Lamps and "Le Parfait" Loudspeaker. French, c. 1925

REGENERATIVE RECEIVER. Signed "The Radio Shop, Sunnyvale, CA". American, c. 1923.

"BRANDY" Point of sale display for C. Brandes headphones. American, c. 1923.

GREBE CR-6. Signed "A.H. Grebe & Co." Short wave regenerative receiver with auxiliary controls and two-stage amplifier. American c. 1920. In August 1919, "Radio Amateur News" featured a cover article by A.H. Grebe. In the article, Grebe described his concep of the auto-radio-phone: "The auto-radio-phone is entirely practical, and the near future should bring extensive developments along these lines, and we may soon hear an SOS: 'Send an emergency service car to car No. 999-999 three miles east of Suburbanville." The photo at lower right of page 139 shows the author in his auto-radio-phone equipped car with a Grebe CR-6 placed in the rea of the vehicle.

Zenith Radio Corporation

LONG DISTANCE RADIO MODEL 1-R. Signed "Chicago Radio laboratory, Chicago, Illinois." walnut cabinet. American, c. 1922.

The Chicago Radio Laboratory began operation in 1918 when two wireless-radio operators set up a "factory" on a kitchen table in Chicago and began making radio equipment for other amateurs. The company operated a small wireless station licensed under the call 9ZN and as word got out, their manufactured products became known as "9ZN Spark Gaps" or "9ZN Receivers."

When the company began advertising in amateur radio magazines, they listed the 9ZN call followed by a small "ith," thus providing the famous trade name Z-Nith. Later the call sign was dropped a the company advertised simply as "Z-nith" (see advertment, right.)

The Zenith Radio Corporation was formed on June 30, 1923 to market Z-Nith products produced by Chicago Radio Laboratory. It was not until several years later that the two merged under the name Zenith Radio Corporation.

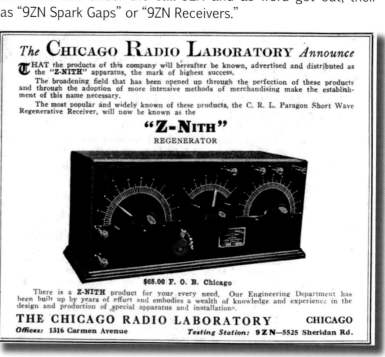

The CHICAGO RADIO LABORATORY *Announce*

THAT the products of this company will hereafter be known, advertised and distributed as the "Z-NITH" apparatus, the mark of highest success.

The broadening field that has been opened up through the perfection of these products and through the adoption of more intensive methods of merchandising make the establishment of this name necessary.

The most popular and widely known of these products, the C. R. L. Paragon Short Wave Regenerative Receiver, will now be known as the

"Z-NITH"
REGENERATOR

$65.00 F. O. B. Chicago

There is a Z-NITH product for your every need. Our Engineering Department has been built up by years of effort and embodies a wealth of knowledge and experience in the design and production of special apparatus and installations.

THE CHICAGO RADIO LABORATORY   CHICAGO

Offices: 1316 Carmen Avenue   *Testing Station:* 9 Z N—5525 Sheridan Rd.

Radio News, 2/1921, p. 564

SUPREME 400B RADIO DIAGNOMETER. Signed "Supreme Instruments Corporation, Greenwood, Mississippi". A fine example of a radio analyzer used by radio repairmen, the device can be used to test radio tubes in and out of the circuit. It is one of the first mutual conductance tube testers and also contains a built in RF oscillator. The meter is capable of measuring AC volts, DC volts, milliamps, and capacitance. American, c. 1929.

2-B TUNING UNIT. Signed "Western Electric." Serial number 251. Used in conjunction with Western Electric's 4-D Superheterodyne receiver to improve selectivity. Brass and aluminum shielded mahogany cabinet supporting ebonite panel measures 11 x 11 x 7". American, c. 1922.

7-A AMPLIFIER. Signed "Western Electric." Amplifier for Western Electric's 10A Loud Speaking Telephone Outfit. Operates with three Western Electric 216A vacuum tubes. Mahogany cabinet supporting ebonite panel measures 12-1/2 x 4-1/2 x 10". American, c. 1922.

RADIO CRAFT RADIOPHONE AMPLIFIER.
Radio-Craft Co. Inc.
American, c. 1920.

RADIO CRAFT RADIOPHONE RECEIVER.
Radio-Craft Co. Inc.
American, c. 1920.

Radio Craft was founded in 1920 by Frank M. Squire, a former draftsman for A.H. Grebe & Co. The company was purchased by De Forest in 1922 as a means to acquire rights to the Armstrong regeneration patent. When Westinghouse notified De Forest that the license was not transferable, De Forest responded by maintaining the company as a separate entity located in a partitioned section of the De Forest factory. The brand was relaunched with a new look, and a new logo closely resembling that of the De Forest Company:

Source: Douglas, Alan. Radio Manufacturers of the 1920's. New York: Vestal Press, 1988.

KEMPER MODEL K-5-2 PORTABLE RADIO RECEIVER. Signed "Kemper Radio Corporation." Five-tube regenerative receiver with loudspeaker, batteries, and antenna contained in a portable leatheroid case.   A fine example of the "all in one" portable design that became popular in the mid-1920s. $16^{1/2}$ x $9^{1/2}$ x $14^{1/2}$" height without antenna, 29" height overall. American, c. 1927

"Portable" radios lead to not-so practical uses

Not nearly as sensitive as modern radios, early receivers required a long antenna and a good ground connection. Typical ground connections were a cold water pipe, a radiator, or even metal box-springs from an old bed, buried in the ground. The antenna usually consisted of a long wire, stretched between trees or along the roof.

By 1923 some sets featured what became known as a "loop antenna," essentially a long antenna wire, wound around a frame. Since all radios made before 1925 operated on batteries, this new "compact" design made portable radios possible. The public wasted no time in coming up with unique applications for this new and exciting technology - some, more practical than others.

During the Leipzig fair this distinctly modernized "sandwich man" paraded the main streets of the town, with a four-tube radio set hanging on his chest and a loop antenna and loudspeaker on his back. Popular Radio, December, 1923

Claimed to be the first wireless equipped delivery truck. Note large horn loudspeaker next to driver. Radio, News June, 1922.

Miss Daisy Crossley tries out her "wireless buoy" at the beach. Wireless Age, October, 1922

"Radio Controlled Car." Radio News, June, 1920.

"Radio on the Ranch." Radio News, February, 1922.

"The Radio Police Car." Radio News, January, 1924.

A family enjoys breakfast in the pool with the cool sounds of their new "portable" radio set. Radio World, August 2, 1924

W. Harold Warren and Guests receiving signals from stations 200 miles distant in a radio roller chair on the Asbury Park, New Jersey, Boardwalk. Wireless Age, August, 1920.

Coast Artillery School advertising truck, equipped with wireless. Radio News, September, 1919

A man operates his automotive radiotelephone device. c. 1922.

A father enjoys a fishing trip with his young son, without missing the big game. Wireless Age, October, 1922.

European radio experimenter Captain Leonard Plugge, of London, with his completely equipped auto, which carries a sensitive nine-tube set capable of picking up all of the European radio stations from all over Europe. The speaker is located in the roof of the car, while the radio controls are located on the dashboard. Radio World, February 5, 1927

The Ski patrol is now equipped with the latest in portable radio devices. Popular Radio, May, 1924

Early example of reporting by radio. Reporters of the Montreal Standard newspaper reporting on the international yacht races on Lake St. Louis. Radio News, September, 1921.

Alfred H. Grebe, founder of A.H. Grebe & Co., in his radio-telephone equipped automobile. Popular Radio, December, 1922

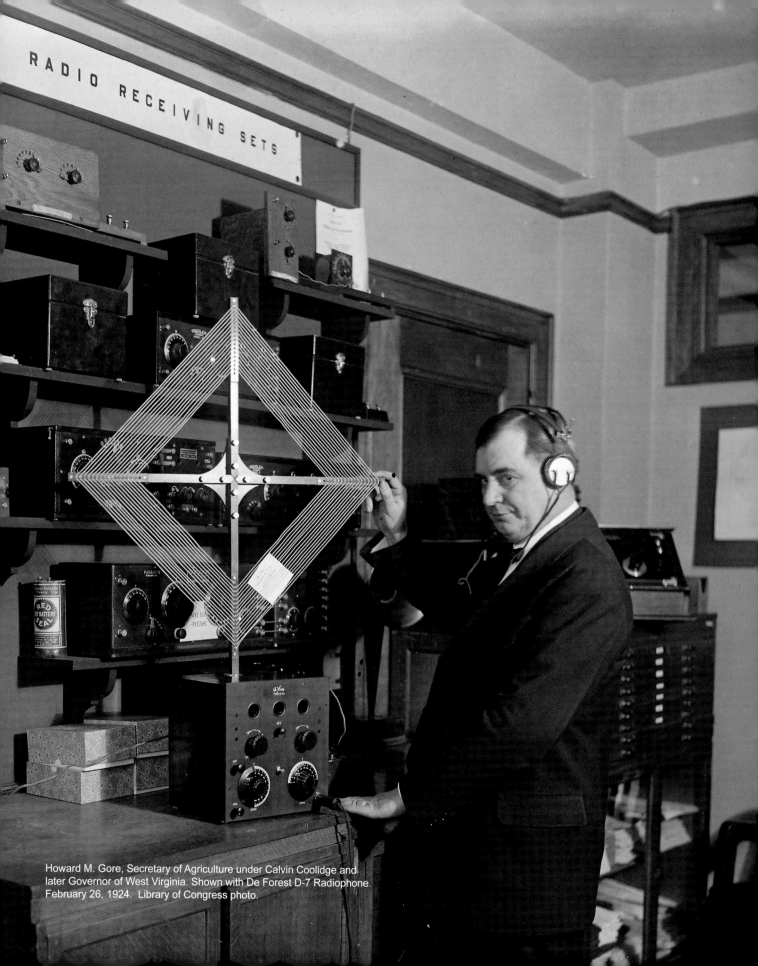

RADIO RECEIVING SETS

Howard M. Gore, Secretary of Agriculture under Calvin Coolidge and later Governor of West Virginia. Shown with De Forest D-7 Radiophone. February 26, 1924. Library of Congress photo.

De Forest Loop Aerial Receivers

DE FOREST D-7A RADIOPHONE. Signed "De Forest Radio Company, Jersey City, N.J." Battery (dry cell) four-tube reflex circuit with crystal detector. Mahogany cabinet 10 x 7 1/2 x 39" including rotatable loop antenna. American, c. 1923.

DE FOREST D-10 RADIOPHONE. Signed "De Forest Radio Company, Jersey City, N.J." Battery (dry cell) four-tube reflex circuit with crystal detector. Mahogany cabinet with shutter-style doors measures 12 1/2 x 9 x 43 1/2" including rotatable loop antenna. Sold for $150 when introduced in July, 1923. American, c. 1923.

DE FOREST D-12 RADIOPHONE (Early version). Signed "De Forest Radio Company, Jersey City, N.J." Battery (dry cell) four-tube reflex circuit with crystal detector. Leatherette cabinet with four-dial front panel with detector adjustment control measures 23 7/8 x 17 x 16" without antenna. Rotatable loop antenna adds 28 1/2" in height. Includes integrated horn loudspeaker.

Not long after the D-12 was introduced, De Forest replaced the crystal detector with a vacuum tube and dubbed the new set the D-17. In one simple move, all of the D-12s sitting on store shelves became obsolete. Faced with irate dealers, De Forest recalled the D-12s and retrofitted them with the new detector. Today, very few of these original crystal-detector D-12s remain. American, c. 1923.

Science and Invention Magazine: December, 1927

RCA VICTOR THEREMIN

*"Music From the Ether"**

Prof. Leon Theremin
demonstrating his invention
c. 1930

The First Electronic Musical Instrument.

Invented by Russian physicist Lev Termen (his name was later changed to Leon Theremin) in 1919, the theremin is the world's first electronic musical instrument.

Eyeing a potential new market, RCA built 500 of them in 1929—but the instrument proved very difficult to play and was discontinued. Today only a handful remain.

Besides looking like no other instrument, the theremin is unique in that it is played without being touched. Two antennae protrude from the theremin - one controlling pitch, and the other controlling volume. As a hand approaches the vertical antenna, the pitch gets higher. Approaching the horizontal antenna makes the volume softer.

Because there is no physical contact with the instrument, playing the theremin requires precise skill and perfect pitch.

Anyone familiar with the horror movies of the 1950s and 60s will recognize the eerie sound of the theremin.

THEREMIN. Signed "Radio-victor Corp. of America." A musical instrument operating entirely on electrical principle, and played by the movement of the hands in space. The instrument covers approximately three and one half octaves, the highest note being about 1400 cycles. The musical note is produced by two oscillators of slightly different frequency beating together. This beat note is then amplified by two audio stages. The change of note or volume caused by the movement of the hand is due to the change of capacity across the respective control rod. Mahogany cabinet measures 19 x 12 x 46-1/2" (not including control rods.) Control rods are 7/16" nickel plated brass tubing, measuring 17-1/2" (Pitch) and 12-1/2" (volume). Eight tubes. American, c. 1929.

* Slogan from RCA Theremin advertising brochure

RADIO RECEIVER MODEL ER-753A. Signed "Manufactured by General Electric Company, U.S.A for Radio Corporation of America." Three-part walnut case designed to open like a book. Single tuning dial controls internal variometer. Perikon detector. Original instruction card glued inside front cover. Sold for $25 at introduction including earphones. American, c. 1922.

RADIOLA SPECIAL. Signed "Radio Corporation of America" but manufactured by Wireless Specialty Apparatus Company. Single tube (UV-199) regenerative receiver. Metal case measures 6 1/2 x 5 1/2 x 10". American, c. 1923.

RADIOLA 60 MODEL AR-954. Signed "Radio Corporation of America." Nine-tube, A.C. powered superheterodyne receiver, also known as the Model AR-954. Two-tone walnut cabinet with bronze escutcheon plate surrounding the tuning dial, supported by bronze-tone moulded feet. Volume control and power switch are the only other front-panel controls. Introduced in August, 1928, the set without loudspeaker sold for $210 including nine Radiotron vacuum tubes. Shown with RCA Model 103 permanent-magnet loudspeaker. American, c. 1928.

RADIOLA 28 MODEL AR-920 RECEIVER. Signed "Radio Corporation of America." Battery powered eight-tube superheterodyne, also identified as the Model AR-920. Dual-tone mahogany cabinet with sloped front measures 27 x 10 $^{1/4}$ x 16 $^{5/8}$" without antenna. Rotatable loop antenna adds 26" in height. Introduced to retail in November, 1925, the set without loudspeaker sold for $210 including eight Radiotron vacuum tubes. American, c. 1925.

RADIOLA 20 MODEL AR-918 RECEIVER. Signed "Radio Corporation of America." Battery powered five-tube Tuned Radio Frequency (TRF) receiver. Dual-tone mahogany cabinet with sloped front measures 19 x 11 $^{3/4}$ x 16". Introduced in November, 1925, the set without loudspeaker sold for $115 including five Radiotron vacuum tubes. American, c. 1925.

RADIOLA VII MODEL AR-905 RECEIVER. Signed "Radio Corporation of America" but manufactured by Wireless Specialty Apparatus Company, also identified as the Model AR-920. This very rare five-tube set didn't perform well and was sold for only a few weeks. Introduced for $290 in November, 1923, complete with Westinghouse FH loudspeaker (shown), tubes, batteries and antenna. Less than 2,000 of these sets were produced, accounting for their scarcity today. Black painted metal front panel supported by mahogany cabinet measuring 24 $^{1/4}$ x 13 $^{1/4}$ x 13 $^{1/2}$". American, c. 1923.

RADIOLA 26. Signed "Radio Corporation of America" but manufactured by Westinghouse. Similar to its sister set, the Radiola Model 24, The Model 26 is completely self contained in a handsome walnut cabinet. Front door includes integrated loop antenna, which can be rotated with open. Removing brass nameplate reveals six dry cell Radiotrons. Scroll-cut Loudspeaker grill reminiscent musical quarter notes conceals a folded horn loudspeaker. Sold for $225 including tubes. American, c. 1925.

RADIOLA 24 MODEL AR-804. Signed "Radio Corporation of America" but manufactured by General Electric. Also known as the Model AR-804. First truly portable Superheterodyne receiver is completely self contained in a black cowhide suitcase-style unit. Removable front cover reveals 2 inner dials, 7 panel speaker plate in gold and black trim, upper cloth grille with vertical bars covers enclosed loudspeaker. Brass scutcheon, headphone jack. Cover also stores detachable, directional loop antenna. Six dry cell Radiotrons with two extra tubes in upper storage compartment. Despite being substantially lighter than the similarly priced Radiola 26, the Model 24 didn't sell well and was discontinued in less than a year. Sold for $195 including tubes. American, c. 1925.

RADIOLA IV. Signed "Radio Corporation of America" but manufactured by General Electric. Also known as the Model AR-880. Three-tube receiver with loudspeaker and batteries enclosed in a handsome walnut cabinet measuring 23 3/4 x 13 1/2 x 14". Introduced in December 1922 for $275, the Radiola IV was the first broadcast receiver made by RCA to have a separate volume control. American, c. 1923.

RADIOLA VI MODEL AR-895. Signed "Radio Corporation of America" but manufactured by General Electric. Also known as the Model AR-895. Consisted of the AR-152 radio-frequency amplifier and AR-1400 detector/amplifier with a faux-mahogany finish dressed up with "pants and a vest" - a mahogany base and matching top. Overall the set measures 23 1/2 x 7 x 11". Introduced in December 1922 for $162.50, without accessories. American, c. 1923.

RCA RADIOLA II, Model AR-800. Signed "Radio Corporation of America." Portable two-tube regenerative receiver packaged in an attractive mahogany cabinet with leather handle. Completely self-contained with built in dry-cell battery storage and headphones. American, c. 1922.

RCA RADIOLA VIIB, Model AR-907. Signed "Radio Corporation of America" but manufactured by Wireless Specialty Corp. Designed as a replacement for the poorly performing Radilola VII. Sold for $275 at introduction in January, 1922, including five vacuum tubes and an integrated loudspeaker. Mahogany case, reminiscent of a table phonograph, measures 23 $^{3/8}$ x 17 $^{1/4}$ x 13" (with lid closed.) American, c. 1924.

RCA RADIOLA 30, Model NS-30-1. Signed "Radio Corporation of America 233 Broadway, New York City" Console featuring inlaid marquetry woodwork with lower front fold-out dial panel including 4 tuning knobs. Superheterodyne receiver. The upper round cloth grille features 5 panel cut-outs and stands on 4 fluted legs. Selling for $575 complete, he Radiola 30 was the first AC-only radio offered by RCA, and one of the first AC powered radios sold in the United States. 8 tubes, battery. Period advertisements include

ATWATER KENT

Radio Receiving Sets and Parts

Complete Receiving Set—Coupled Circuit Tuner and Detector 1-stage Amplifier
A similar set is furnished with Detector 2-stage Amplifier

Mounted Variometer

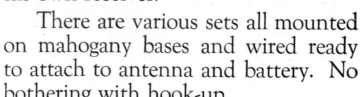

Mounted Variocoupler

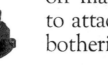

Detector Unit

ATWATER KENT products sell on appearance; they stay sold on quality of performance.

This is the reason for the popularity of ATWATER KENT Radio Equipment.

Look over the illustrations. They show a portion of the line, which includes complete sets, as well as parts from which the radio fan may build his own receiver.

There are various sets all mounted on mahogany bases and wired ready to attach to antenna and battery. No bothering with hook-up.

Use a set as YOUR demonstrator

Detector 1-stage Amplifier
A similar unit is furnished
in a 2-stage Amplifier

Detector 2-stage Amplifier

Standard Tube Socket

1½-Volt Tube Socket

ATWATER KENT MANUFACTURING COMPANY, PHILADELPHIA, PA.
Radio Department   4947 STENTON AVE.   *Write for Literature*

ATWATER KENT

Radio Receiving Sets and Parts

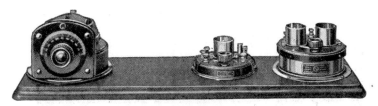

Complete Receiving Set—Coupled Circuit Tuner, Detector Unit and 2-stage Amplifier
This Set is also furnished without Amplifier

THESE sets are attractive enough in appearance to be installed in the most "exclusive" home or club; moderate enough in price to be within reach of anyone; and excellent enough in operating qualities to satisfy the most particular.

For the fan who wants to experiment with varying hook-ups there are parts which will meet every requirement.

ATWATER KENT Radio Sets and Units sell readily and are a very profitable line.

They're always SEEN in the show windows

Unmounted Variometer

Unmounted Variocoupler

Condensite Dials
in 50 or 100 pt. graduations
for $\frac{3}{16}$" or $\frac{1}{4}$" shafts

Type L Transformer A.F.

Transformer A.F.

Panel Rheostat

Table Rheostat

ATWATER KENT MANUFACTURING COMPANY, PHILADELPHIA, PA.
Radio Department   4947 STENTON AVE.   *Write for literature*

ATWATER KENT NUMBER 4333 "COMPACT" RECEIVER. Marketed as the Model 5, this set was introduced in September of 1923 and is virtually identical electrically to its predecessor, the Model 4066. It is one of the first examples of a "compact" design, with most of the components packed tightly into the 5-tube base.   Unfortunately about the time the radio was introduced, the FCC expanded broadcasting frequencies.   This greatly expanded the number of stations on the dial and placed higher demands on radio receivers to sort them out. The unit didn't perform well and was soon discontinued. Today the Atwater Kent Model 5 is quite rare and is highly prized by vintage radio collectors. American, c. 1923.

Atwater Kent Open Receiving Sets

Atwater Kent, a successful manufacturer of automotive electrical parts, introduced a line of radio components in 1921 consisting of modular assemblies that could be connected together and mounted on a wooden board. Recognizing the popularity of these "breadboard" receivers and the rapid adoption of radios beyond the hobby market, Kent introduced a line of fully assembled "Open Receiving Sets" beginning in November of 1922. Today these unusual sets are among the most prized by collectors.

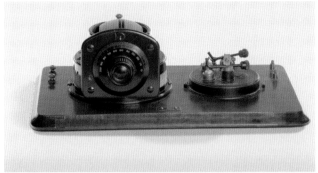

CRYSTAL RADIO RECEIVER. Atwater Kent never sold a completely assembled crystal set. The set shown above was assembled from individual components marketed by the company in the early 1920s. Mahogany board measures $9^{1/2}$ x 16". American, c. 1923

NUMBER 3925 REGENERATIVE RECEIVER (Model 1). The Model 3925 was introduced late in 1922 and consisted of a tuner, detector, and 1 stage amplifier. In order to avoid payment of patent fees for use of a regenerative circuit the unit was sold without the middle variometer installed. Kent sold the variometer as a separate item (frequently in the same advertisement as the radio, although its real function was never mentioned) and when installed by the user it provided additional RF signal boost through regeneration. Mahogany board measures $8^{1/4}$ x $18^{1/2}$". American, c. 1922

NUMBER 3945 REGENERATIVE RECEIVER (Model 2). Atwater Kent offered an additional stage of audio amplification with the 3-tube Model 3945. It was shipped about the same time as the Model 1. Mahogany board measures 8 x $25^{3/4}$". American, c. 1923

NUMBER 3955 REGENERATIVE RECEIVER (Model 3). In the Model 3955, Atwater Kent's designers separated the detector tube in order to produce a clearer signal, free of interference from the audio amplifier. This set was sold with only the coupled circuit tuner, detector and 2-stage amplifier installed. Additional components were added by the dealer or end user. Mahogany board measures $8^{1/2}$ x $24^{1/4}$". American, c. 1923

NUMBER 3955 RECEIVER.. This is the same basic set as the Model 3 above, but with an RF amplification stage instead of a variometer, improving sensitivity without making the set regenerative. Mahogany board measures $8^{1/2}$ x $24^{1/4}$". American, c. 1923

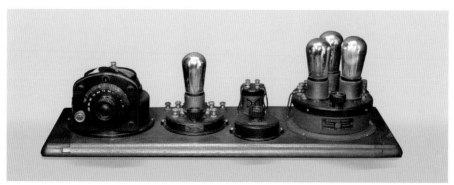

NUMBER 4052 RECEIVER (Model 6). In order to increase sensitivity, Atwater Kent added an RF amplifier to the existing model 3945, and the Coupled Circuit Tuner was replaced with a Type 11 Tuner. American, c. 1923.

NUMBER 4066 RECEIVER WITH RF AMPLIFICATION (Model 7). Introduced in early 1923, the 4066 added another stage of RF amplification, eliminating the need for a regenerative circuit. This was the first five-tube radio produced by the Atwater Kent factory, and was the first AK set offered for sale in an optional cabinet, although none of the cabinets survive today. Mahogany board measures $8^{1/4}$ x $26^{1/4}$". American, c. 1923.

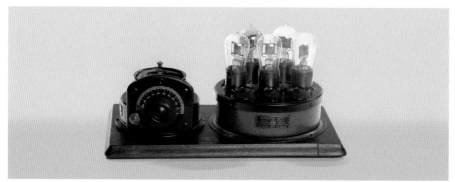

MODEL 5 "COMPACT" RECEIVER (Number 4333). The Model 5 was the first of the open sets marketed with a model number. It was introduced about nine months after the Model 4066 and is electrically virtually identical to its predecessor. It is one of the first examples of a "compact" design, with most of the components packed tightly into the 5-tube base. Unfortunately about the time the radio was introduced, the FCC expanded broadcasting frequencies. This greatly expanded the number of stations on the dial and placed higher demands on radio receivers to sort them out. Unfortunately the Model 5 didn't perform well and was discontinued. Today the Atwater Kent Model 5 is quite rare and is highly prized by vintage radio collectors. Mahogany board measures $9^{1/4}$ x 16". American, c. 1923

NUMBER 4445 RECEIVER (Model 9). The Model 9 was introduced shortly after the Model 5, returning to a single stage of RF amplification, but with both a Coupled Circuit Tuner and and Type 11 Tuner. This earlier Model 9 features terminal posts instead of a battery cable, and a green TA Unit (The three-tube assembly.) Mahogany board measures 8¹/² x 26¹/²". American, c. 1923.

NUMBER 4445 RECEIVER (Model 9, later version). Introduced about a year after the original Model 4445, this later version features brown crinkle paint on the TA Unit, and power for the set is carried through a cord. Mahogany board measures 8¹/⁴ x 24¹/²". American, c. 1923.

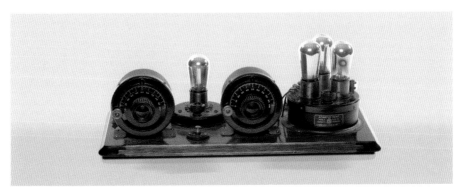

NUMBER 4660 RECEIVER (Model 9C). In April of 1924, Atwater Kent introduced the Model 4660, which replaced both inductive tuners with variable condensers. Electrically, the Model 9C is identical to a Model 10C, but without one RF amplification stage. Mahogany board measures 9³/⁴ x 20¹/⁴". American, c. 1924.

NUMBER 4340 "RADIODYNE" RECEIVER (Model 10) The Radiodyne was introduced in the fall of 1923, employed two stages of RF amplification and was the first successful AK radio employing condenser tuning (The Model 8, introduced at the same time as the Model 5 also used a single tuning condenser, but it did not sell well.) The Radiodyne was built in order to capitalize on the popularity of the new 3-dial, 5-tube TRF Neutrodyne radios such as the FADA 160 and Freed Eisemann NR-5. Unfortunately the "Radiodyne" trademark was already held by Western Coil, so Atwater Kent was forced to discontinue the unit. The Radiodyne is distinguished by its green tuning condenser cans and the "Radiodyne" label on the transformer tops. Note: The model shown in the photo is stamped with production number 3216 on the bottom of the board. This confirms that it is in fact a Radiodyne. The variable capacitors (which are not Radiodyne) were likely replaced in a factory modification done to reduce a capacitance effect. Mahogany board measures 10 x 31$^{1/2}$". American, c. 1923.

NUMBER 4550 RECEIVER (Model 10A). The Model 10A was introduced to replace the Radiodyne. It can be distinguished from the other Model 10's by its vertical RFcoils, the presence of a separate potentiometer located between the first and second stage, and the lack of a switch on the first RF coil. The Model 10A suffered from a tendency to oscillate, partly due to the vertical orientation of the RF transformers.   A modification was made to orient the coils sideways with the resulting set receiving the Model 10B designation (below). Mahogany board measures 10 x 31$^{1/4}$". American, c. 1924.

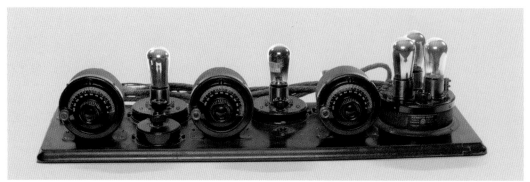

NUMBER 4550 RECEIVER (Model 10B). The Model 10B is recognized by the orthogonal (sideways mount) RF coils, the presence of a separate potentiometer located between the first and second stage, and the presence of a switch on the first RF coil. Mahogany board measures 10 x 31". American, c. 1924.

NUMBER 4700 RECEIVER (Model 10C). The 10C also has orthogonal (sideways mount) RF coils, but the separate potentiometer has been removed and replaced with fixed damper resistors hidden beneath the board. In addition, the diameter of the tube base for the second tube has been significantly reduced by elimination of the filament rheostat.   These changes made a shorter board possible, reducing the overall length by two inches. Mahogany board measures 10 x 29$^{1/4}$". American, c. 1924.

NUMBER 4910 RECEIVER (Model 12). In early 1924, Atwater Kent added an additional stage of amplification to the Model 10 to create the 6-tube Model 12, the last of the "breadboard" line (well, almost the last. About the same time they also shipped the 9C, essentially the same set as the Model 9 but with condensers in place of the variometers.) Mahogany board measures 10 x 33$^{3/4}$". American, c. 1924.

Source: Douglas, Alan. Radio Manufacturers of the 1920's. New York: Vestal Press, 1988.
Williams, Ralph O., The Atwater Kent Radios. The AWA Review. 12 (1999): 27-80.

CLOWN CRYSTAL RADIO RECEIVER. Signed "RADIOGUS".
Wood clown resting on arms, Galena crystal mounted on
clown's head, cat-whisker reaches down from toes. Tuning
is accomplished by opening and closing the clown's coat,
adjusting a variable inductance hidden within. Measures
5³/₄ x 6¹/₂ x 13¹/₂". French, c. 1920s

Crystal Radios

At the turn of the 20th century, an American scientist, Greenleaf Whittier Pickard, found that a number of naturally occurring crystalline minerals could be used to detect radio signals. Radios employing this kind of detector became known as crystal radios. In the typical early radio-wave crystal detector, the crystal rock was fixed into a brass cup and the radio operator found the loudest signal by touching the wire, called a cat's whisker, to various points on the surface of the crystal.

In the early days of radio, people built and used simple and inexpensive crystal radio sets that worked without electrical power from wall sockets or batteries, and this technology was known as wireless. Even after vacuum-tube radios came into widespread use following World War I, crystal radios remained popular, especially among beginning amateur radio enthusiasts, boy scouts, and school kids, who continued to build crystal radios as their introduction to the field of communications.

During the Great Depression, a perfectly workable crystal radio detector could be constructed from a five-cent piece of galena crystal and the wire from a safety pin, and building and using homemade crystal sets brought endless hours of enjoyment to children of the Great Depression. After the detector was connected to iron bedsprings (which doubled as an antenna) and grounded to household cold-water pipes, a youngster needed only inexpensive headphones to bring in the world of radio—all the power needed to run the crystal set came from the 'air.' Later, GIs of World War II constructed similar wireless radios from rusty razor blades and pencil lead, the iron oxide crystals of the rust replacing the galena crystal and the graphite of the pencil lead substituting for the safety-pin 'wire.' These crystal radios were known as foxhole radios.

Over the years, fascination with crystal radio building has never died, perhaps because the technical achievements of the communication media cannot dim the enchanting simplicity of the crystal radio in its demonstration of the wonder that radio really is.

Text courtesy of PV Scientific Instruments, www.arcsandsparks.com

PERIKON CRYSTAL DETECTOR. Signed "WIRELESS SPECIALTY APPARATUS CO. NEW YORK SERIAL NO. P.R.C. 100". Ebonite panel on Mahogany base, supporting Nickel plated detector, battery switch and six binding posts. Engraving on side of panel reads "MODIFIED TYPE 1503". Perikon is an acronym for "perfect Pickard contact." American, c. 1910

PYRON CRYSTAL DETECTOR. Signed "WIRELESS SPECIALTY APPARATUS CO. 81 NEW ST. NEW YORK SERIAL NO. P.R.C. 100". Ebonite panel on Mahogany base, supporting Nickel plated detector, selector switch and FIVE binding posts. American, c. 1910

Pickard worked for the American Telephone and Telegraph Company from 1902 to 1906. While reading an article in the London Science Abstracts he got an idea and tested a large number of minerals in an effort to discover the most effective detector of radio waves. After about eight months he determined that silicon produced excellent results, and he received a patent on a silicon crystal detector in 1906. Reportedly, he tested more than 30,000 combinations of materials for detectors.

In 1907, Pickard joined with Phillip Farnsworth and John Firth to form the Wireless Specialty Apparatus Company to market his patented detectors, one of which was called Perikon, an acronym for "perfect Pickard contact." Two of these early detectors (shown above) are on display at the museum.

Wireless Specialty Apparatus Co, went on to become one of the major suppliers of wireless equipment for military and commercial use.

PHILMORE SELECTIVE . Signed "Philmore Mfg. Co., New York." Cathedral style folded sheet metal cabinet measures 6 x 5 x7". American, c. 1932

FEDERAL JR. Signed "Federal Tel. & Tel. Co." Brass case Measures 8-1/2 x 5-1/2 x 6-1/2". The cases are actually old Ericsson telephone cases purchased after the Ericsson factory closed. American, c. 1922.

THE CRYSTAL DYNE. Signed "Bethlehem Radio Corp". Unusual stamped steel case measures 8-1/2 x 3-7/8 x 2-1/8". American, c. 1929

SPCO CRYSTAL RECEIVER. Signed "Steel Products Corp. of California, San Francisco, CA." Stamped steel case measures 5 x 4 x 2". American

RECEIVING SET TYPE RD. Signed "Adams-Morgan Co." Mahogany cabinet measures 15-1/2 x 11-1/2 x 9-1/2 (height includes loose coupler.) American, c. 1912

G. PERICAUD OUDIN. This set was listed in the Pericaud catalog with the following comment: "This station allowed the reception of telephone emissions and concerts from the Eiffel Tower..." French, c. 1919 - 1923.

BROWNIE. Signed "J.W.B. Wireless, Co." Cardboard tube supporting ebonite top panel. Measures 5 x 3-1/4". British, c. 1923.

JUBILEE. Signed "Jubilee Mfg. Co., Omaha, Neb." Measures 5-1/4 x 4-1/4 x 7/8" at base; coil is 2-1/2" dia., 3-1/8" in length. Marketed as "The best buy in radio." American, c. 1926.

"THE SERVICE." Signed "Edison Bell Ltd." Oak box with hinged lid, supporting ebonite panel and nickel plated hardware. Nearly identical to the Edison Bell Type B, but without BBC stamp. Measures 8-3/4 x 5-3/4 x 7". British, c. 1923

CRYSTAL RADIO RECEIVER. Signed Martian Mfg. Co. Designated "Martian Big Four." Despite its spaceship-like appearance, the set was not named for inhabitants of the planet Mars, but for C.L. Marti, the founder of the Martin Mfg. Co. Overall height is 7-1/2". American, c. 1923

"UNCLE TOM" CRYSTAL RECEIVER. Unsigned but by Grafton China Works. Painted ceramic with coil wound around top-hat with slide-tuner in rear. Bow tie cat's whisker and crystal as "stud" on shirt on oval base with wiring hidden inside body. Standing 9" tall. British, c. 1924.

BROWNIE #2 CRYSTAL RADIO RECEIVER WITH "TWO STAGE NOTE MAGNIFIER." Signed "Brownie Wireless Co." When introduced in 1923, this unusual and interesting radio was designed to be used with headphones.

In 1926, a two-valve ("valve" being the British term for vacuum tube) amplifier was added so that the set could be used with a loudspeaker. The amplifier was cleverly designed to cradle the radio from beneath. British, c. 1923 (receiver) and 1926 (note magnifier).

RADIO RECEIVER No. 797. Signed "Hunt & McCree, New York." Oak base measures 7-1/2 x 4-1/2 x 3/4" incl. feet. This is one of the earliest commercial crystal sets and is the oldest crystal radio at the Museum. American, c. 1912

THE KENMAC BOOK. Signed "Kenmac Radio Ltd." Disguised to look like a pocket book. Green leather "binding" measures 4-3/4 x 3-1/2 x 1". British, c. 1925

MINIATURE CRYSTAL RADIO. Unsigned. Measures 3-1/4 x 2-5/8 x 2". Shown in photo with an American dollar for scale. c. 1922

THE ZENITH WINCHARGER

Two brothers from Cherokee, Iowa, John and Gerhard Albers, originally developed the famous 6-volt wind generator in 1927. Its popularity spread throughout the the rural mid-west where electricity was largely unavailable. Zenith Corporation purchased Wincharger in the mid 1930s and soon developed 12-volt, 32-volt, and 110-volt generators as well. By the late 1930s the Zenith company was known all over the world for the unfailing dependability of the Wincharger.

By 1950 power lines were extended virtually everywhere, with energy becoming available and inexpensive. Several power companies refused to service a farm with a functioning Wincharger and soon the dependable generator faded into obscurity.

Though the original is no longer manufactured, people still maintain and use Winchargers as an alternative energy source. Wincharger is experiencing a rebirth and a variety of industrious individuals continue to manufacture and advertise replacement blades and parts.

Shown here with Museum staff member Tana Granack, the Wincharger measures 107$^{1/2}$" in total height, and 60" in total width. The single wood vane is 72$^{1/2}$" long, and the angle-iron stand reaches 72" tall. American, c. 1930s.

Cathedral Loudspeaker
Pathé Phonograph and Radio Corp.
American, c. 1927

S.G. Brown, LTD
Type Q
British, c.1924

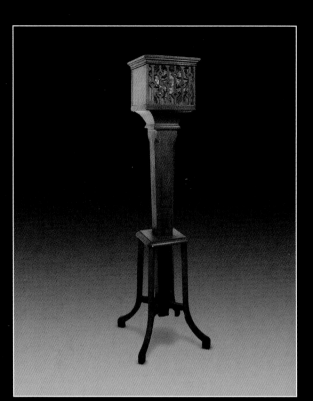

Saal Pedistal
H.G. Saal Co.
American, c. 1925

Sea-Tone Loudspeaker
Tonks Brothers, Newark, NJ
Made with an authentic seashell.
American, c. 1925

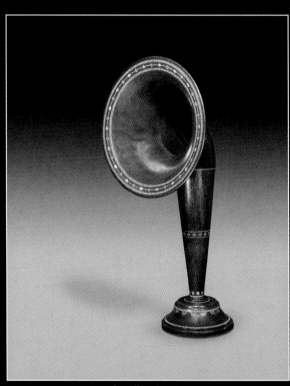

Sarnoff Loudspeaker
Hand-made in Czechoslovakia for
RCA Chairman David Sarnoff.
Mahogany with Mother-of-pearl inlay
c. 1924

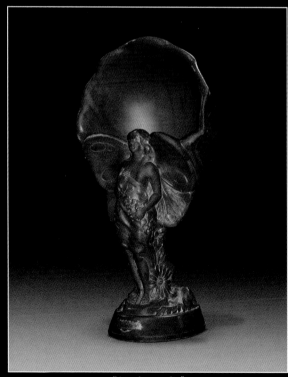

"Voice from the Sky"
Florentine Art Productions
American, c. 1925.

"Miss Muffet" by Artandia Ltd.
British, c. 1927.

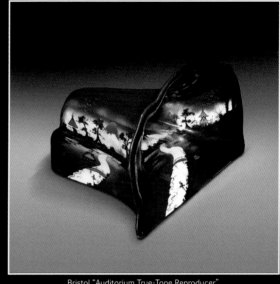

Bristol "Auditorium True-Tone Reproducer"
British, c. 1926.

"Rose Bowl"
British Electrical Manufacturing Co.
British, c. 1926.

"Persian King"
Doulton & Co. for Artandia Ltd.
British, c. 1927.

"The Endeavor" by Jodra Mfg. Co.
American, c. 1927.

Utah Radio Products Co. Model 30
American, c. 1926.

"Radiolux"
Amplion Corp. of America
American, c. 1925.

"The Selector"
Selector Co.
American c. 1927.

S.G. Brown, LTD
Crystavox
(British)
1923

S.G. Brown, LTD
SER H2
(British)
1923

S.G. Brown, LTD
Type H1
(British)
1920

Sterling Baby
(British)
1924

S.G. Brown, LTD
Type Q
(British)
1924

TMZ Junior
Telephone Mfg. Co.
(British)
c. 1922

S.G. Brown, LTD
Silver Hunting Horn
(British)
1924

True Music Junior
GEC Gecophone
(British)
1922

Atwater Kent Type M
(American)
1924

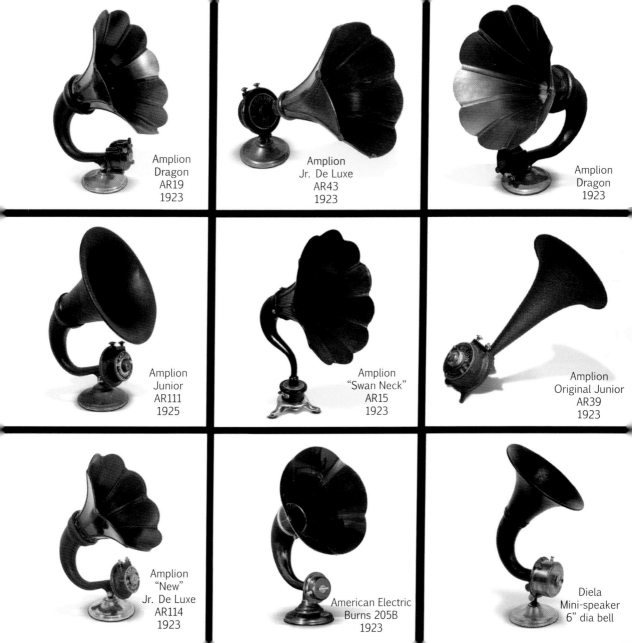

Amplion
Dragon
AR19
1923

Amplion
Jr. De Luxe
AR43
1923

Amplion
Dragon
1923

Amplion
Junior
AR111
1925

Amplion
"Swan Neck"
AR15
1923

Amplion
Original Junior
AR39
1923

Amplion
"New"
Jr. De Luxe
AR114
1923

American Electric
Burns 205B
1923

Diela
Mini-speaker
6" dia bell

GALLERY FIVE
The Golden Age of Radio

The Golden Age of Radio

Improvements in radio mark an important milestone in four centuries of electronic discovery. Radio is now the centerpiece of the family living room and profoundly impacts American life during the hard times of The Great Depression and WWII.

Eager home audiences lead to rapid growth and unprecedented influence for the advertising and entertainment industries.

At the peak of its popularity during these dark years, radio is the lifeblood of the country. Radio programs not only entertain and inform, they strengthen bonds between neighbors, communities and nations.

Do you think the early inventors ever imagined the world of radio?

The information age is born!

GALLERY FIVE
THE GOLDEN AGE OF RADIO

IN 1920, WHEN KDKA WAS LAUNCHED, there were an estimated five thousand radio sets in this the United States. Two years after KDKA began broadcasting, the number of stations jumped to over five hundred. By 1924, there were over three million radio sets and over one third of all money spent on furniture in the U.S.A. went to buy radios. No longer were people confined to small black boxes with wires and tubes and headphones, but now owned radios encased in ornate wooden boxes that they could proudly display in their living rooms.

During the peak of radio production, there were literally hundreds of manufacturers of fine radios, many of which were ornate showpieces for the home. The Museum's collection includes a broad spectrum of beautiful radios from this period including Atwater Kent, Crosley, RCA, Zenith, and many more.

Quite a few American radio manufacturers went out of business during the Great Depression. The longest surviving American manufacturer of radios was Zenith. Sadly, it too is no longer in business.

A few radios from the Golden Age

When the easy-to-use Superheterodyne radios became popular, demand arose for more and better programming. During the depression, people became so addicted to their radios that their radio set was last thing they would sell. As the public's appetite for better programming grew, the only way to satisfy this demand was through the larger budget a network could provide. Networks, unlike independent local stations, could afford to hire big-name stars, while

The Museum's 1920s radio "store"

the independents were confined to reporting sports events and local news, playing records, and producing low budget drama.

In 1926, David Sarnoff founded NBC, the world's first broadcasting network, and was ruthless in his expansion of the company. "I don't get ulcers," he is quoted as saying, "I give them." NBC became famous by broadcasting Charles Lindberg's homecoming after his successful cross-Atlantic flight. Sarnoff's network gained many followers by virtue of its outstanding programming.

Under Sarnoff's direction, NBC later focused heavily on introducing television to the commercial world. In 1930, Sarnoff became president of RCA. In 1933, President Roosevelt broadcast his first fireside chat, thereby transforming the radio into a political tool. By then, this medium's entertainment value already was well established. Whether one lived in the city or on farm miles from one's neighbors, the radio brought entertainment into the home.

At first, NBC actually consisted of four associated networks. The largest two were called the Red Network and the Blue Network. In 1927, a competing network, the Columbia Broadcasting System ("CBS"), was formed by a disgruntled talent agent whose clients were rebuffed by David Sarnoff. Later, the Federal Communications Commission forced NBC to divest itself of the Blue Network, which thereafter became the American Broadcasting System, or "ABC." There were other radio networks as well, including the Mutual Broadcasting System. Today there are numerous radio networks, but of the early ones, only NBC, ABC and CBS

Franklin D. Roosevelt (1882 - 1945)

made the jump to television.

Content of the early networks covered a wide range of subjects and fostered the creation of thousands of new public personalities. One of the most popular programs was Amos and Andy. The Lone Ranger was another weekly series that appealed to children as well. Classical music was popular, and NBC had its own symphony orchestra.

Many stars were born during the Golden Age of Radio, including vaudeville greats George and Gracie Burns, Bob Hope, Jack Benny, Ed Wynn, Red Skelton, and Jimmy Durante to name just a few. Singing greats included Rudy Vallee, Gene Autry, Kate Smith, Bing Crosby and Frank Sinatra. Big Bands such as Tommy Dorsey, Duke Ellington, Benny Goodman and Glen Miller established their national reputations through radio. Careers were started from a single broadcast, such as Orson Welles and his rendering of the War of the Worlds. Radio personalities, such as Lowell Thomas, Walter Winchell, Jack Haley, Charlie McCarthy, Arthur Godfrey and Ronald Reagan, became household names. Not until the 1960s and 70s was a comparable outpouring of musical talent seen, one that arguably pales in comparison to the Golden Age of Radio.

News correspondents such as Walter Cronkite became famous for their broadcasts from Europe during World War II. Radio sets at that time included a shortwave band to enable listeners to keep track of the war's progression by receiving broadcasts from Europe. Radio brought Franklin Delano Roosevelt into people's homes, and many people hung a portrait of the president next to the family radio.

Radio's effect on society was immeasurable. Today, some people complain about the role of television in society, but they have forgotten that in its day, the effect of radio was just as dramatic. Prior to radio, people often spent their evenings socializing with friends or neighbors, listening to phonographs, player pianos, music boxes, or having a sing-along. Several fine examples of these pre-radio entertainment devices can be found on page 53.

With the advent of radio, people were able to bring their entertainment right into their own homes, and the need to go visiting decreased dramatically. Until television became popular in the 1950s, the country was obsessed with radio. The explosion of radio talent certainly had its roots in a technological development, but it was more than that. Man had just tamed electricity, flown across the Atlantic, developed massive steel factories, built skyscrapers, and motorized the average family so that they could travel anyplace they chose. The radio allowed everyone to be part of a much larger wave that was transforming the entire world.

Until the recent advent of Satellite Radio, there have been only two major innovations in radio receivers since the 1930s. The first took place in the 1950s, when vacuum tubes were replaced by transistors, resulting in an explosion of portable radios that could fit in a shirt pocket. The other major development was the FM band, invented by Edwin Armstrong in 1933. By the 1960s, FM had finally replaced AM as the format of choice in radio.

Although radio broadcasting does not command the dominant position it once did, it still is an integral part of our lives. Recent innovations such as talk radio, satellite, HD radio, and sideband broadcasts of everything from traffic to stock prices demonstrate a vibrant market, and there appears to be no end in sight.

MAGIC-TONE NOVELTY RADIO, Model i504, Table-N. Signed "Produced, bottled and blended by Radio Development and Research Corp., Jersey City 4 N. J." Amber colored glass liquor bottle shape, dial on bottle neck, cap is on/off/volume control, superhet, broadcast, medium wave, 10 volt radio, 4 tube AC/DC. Label reads "100% Proof performance," and "The radio that soothes your spirits." American, 1947

LYRIC FLOOR CABINET RADIO. Signed "Rauland Corp, Chicago, Illinois & All-American Mohawk Corp., Chicago."
Six legged Chinese-style, Chippendale console, featuring vivid hand painted images on carved wood double-door cabinet. Lyric Electro-dynamic 12 inch speaker, set behind ornate, scroll cut-out cloth speaker grille.

Early AC set, originally built as a battery powered radio and then exchanged and equipped with AC plug
(Until 1925 all radios were battery powered.) American, c. 1928.

CROSLEY CATHEDRAL TABLETOP RECEIVER, Model 148. Signed "Crosley Radio Corp., Cincinnati, Ohio." Wood case, cloth grille with cut-outs, 3 tuning knobs, 5 tube, superheterodyne. Crosley offered their popular "Fiver" circuit in several different cabinet styles and finishes. The model 148 demonstrated Crosley's ability to provide good performance at a reasonable price. The cathedral style was originally dubbed a "beehive set." Collectors later adapted "cathedral" as it is more descriptive of the radio's appearance. American, 1936

PHILCO MODEL 20 "BABY GRAND" RECEIVER. Signed "PHILCO CORPORATION." Walnut cathedral style cabinet, metal center front escutcheon, cloth loudspeaker grill with scrolled cut-outs. This set established Philco as the leader in broadcast radio receivers for the entire decade of the 1930s. The set featured an electro-dynamic speaker and an 7-tube receiver in a compact table-top cabinet, selling for the unheard of price of $49.50 (without tubes). The Model 20 was extensively advertised by Philco, and over 300,000 sets were sold. American, c. 1930.

RCA 'RADIOVOX'. Signed "Radio Corporation of America, Manufactured by Gilfillan Bros. Inc." Cathedral, wood cabinet, center front window dial with "Radiovox" printed on the metal escutcheon, 3 knobs, upper cloth grille with ornate rounded cut outs, decorative left and right horizontal fluting, 6 tubes, TRF, broadcast only, AC. Very rare. American, c. 1928

CROSLEY 'LITLFELLA' CATHEDRAL, Model 125. Signed "Crosley Radio Corporation, Cincinnati, Ohio" Early Superheterodyne tabletop with 8 inch Magnavox speaker. Upper grille features 3 cutouts covered with original gold grille cloth. Center front window dial surrounded with molded press-wood escutcheon (made to look like brass), fluted columns, 3 tuning knobs, 5 tubes, AC. Inspired by Philco's "Baby Grand" cathedral radio, Crosley produced this well made and more affordable set. American, c. 1932.

CROSLEY 'BUDDY BOY' Model 58. Signed "Crosley Radio Corp., Cincinnati, Ohio." Cathedral tabletop, ornate repwood cabinet, thumbwheel tuning, cloth grille with cut-outs, two knobs, AM broadcast, shortwave, AC. Part of a line of Crosley radios featuring partial or complete repwood cases. Repwood was a unique combination of wood fiber and plaster recreating the look of finely carved wood. Extremely heavy. American, c. 1931

CROSLEY 'SHOW BOY' Model 59. Signed "Crosley Radio Corp., Cincinnati, Ohio." Cathedral with ornate repwood case, 3 knobs, center front window dial, upper grille with 4 section cloth cut-outs; 5 tubes BC. Repwood was popular in the 1930s until the cost and weight eventually made the material impractical. American, c. 1931

SPARTON 'SELECTIME' CLOCK RADIO Model 738 Signed "Sparks-Withington Company, Jackson, Michigan" Tombstone featuring rich two-tone wood with 2 lower center front knobs and automatic/manual toggle switch. "Telechron" clock, white face, surrounded by 48 switches which enable the listener to automatically control on and off power to the radio in a 12 hour cycle. Upper-middle center escutcheon featuring 6 push-button station selector; Upper cloth grille featuring 3 vertical bars, 8 tube chassis, superheterodyne, broadcast only, AC, American, c. 1935.

ADVANCE ELECTRIC COMPANY 'THE FALCK' "Licensed by and manufactured under supervision of Advance Electric Company, Los Angeles, California." Table model, wood/walnut case, 2 knobs, metal escutcheon surrounding right center dial, reads "Falck" above tuning knob. Upper cloth grille with scalloped cut-outs emulating rays; 6 tubes, TRF circuit, toggle power switch on the back, standard broadcast, AC. Improvement over original "Cub" model (1930) with advanced circuitry and simpler component layout. American, c. 1931.

RCA RADIOLA 47. Signed "Radio-victor Corporation of America, Chicago, Ill."
Lowboy console radio with walnut cabinet, featuring single control cluster for tuning and volume control with dial screen located above. On/Off toggle switch located on the side. Features include "local-distant" switch for "two stage radio frequency amplification." Broadcast range from 550 to 1500 kilocycles. Manual 78 rpm electric turntable and electric amplification, 5 tubes. Original instruction booklet reads "Lighting circuit operated, antenna type, shielded radio receiver combined with an electric phonograph featuring RCA electro-dynamic loudspeaker." Four carved legs. Early A.C. set. American, c. 1929.

WESTERN AIR PATROL, Model 28. Signed, "Western Auto Supply, Los Angeles, CA." Rare superheterodyne tabletop radio. Two-toned, wood cabinet. Two tuning knobs located below separate viewing windows; center cloth grille featuring 5 panel wood cutout. Unique AM and tuning switch located in back. SW/BC, 5 tubes. American, c. 1936

KADETTE "JEWEL" Model 434964. Signed "Licensed and manufactured by International Radio Corporation, Ann Arbor, Michigan." Cabinet is walnut colored Bakelite with ornate beige-colored center grille and plastic cut-outs. Right front dial, 2 knobs, BC, 3 tubes, AC/DC. American, c. 1937

EMERSON "DUO-VOX" AC-DC RECEIVER, Model 107
Beautiful hand-rubbed matched American butt walnut with mahogany inlay, front and back features use of new metal tubes and "telegraphic interference trap". From a 1936 Emerson Radio advertisement: "The Model 107 has the same effect on the eye as on the ear; simply breathtaking!" Seven tubes. American, c. 1937

By the late 1930s, taller table radios like the classic cathedral or tombstone cases were out of fashion, and a new style, the 'laydown' (see below) was introduced. Instead of putting the speaker above the chassis, the manufacturers laid it over and put the speaker on the side.

ZENITH TABLE RADIO, Model 6-S-528, Chassis 6A02. Signed "Zenith Radio Corporation, Chicago, Illinois." Wood burl veneer with right front (white) dial, plastic escutcheon featuring 5 push-button tuning, left vertical grille bars, three knobs, 6 tubes, AM standard broadcast, shortwave, police/amateur, AC. An attempt to emulate the sophisticated cabinet design of a more expensive competitor (i.e. Emerson). American, c. 1941

FIRESTONE AIR CHIEF, Model S7403-6. Signed "Firestone Tire and Rubber Company, Akron, Ohio." Table top, wood, 3 knobs, Ingraham cabinet, vertical grille, superheterodyne, shortwave, approximately 19 inches wide. The renowned Ingraham Company, who designed the case, was famous for crafting fine clocks and compound curve cabinets. American, c. 1940

PILOT 'MIDGET' RECEIVER. Signed "Pilot Electrical Manufacturing Co., NY" Art deco tombstone design with squared "Metropolis" inspired cabinet. Two matching carved wood tuning knobs, 2 identical vertical inlaid designs on cabinet, center tuning window, upper cut-out grille cloth; 6 tubes. Pilot Electrical was already successfully manufacturing batteries and parts before expanding into radios. American, c. 1932

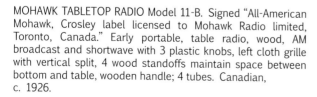

MOHAWK TABLETOP RADIO Model 11-B. Signed "All-American Mohawk, Crosley label licensed to Mohawk Radio limited, Toronto, Canada." Early portable, table radio, wood, AM broadcast and shortwave with 3 plastic knobs, left cloth grille with vertical split, 4 wood standoffs maintain space between bottom and table, wooden handle; 4 tubes. Canadian, c. 1926.

WESTINGHOUSE TABLE RADIO, Model WR272L. Signed "Westinghouse Electric Corp,. Home Radio Division, Sunbury, Pennsylvania." Wood case with right front slide-rule dial, left wrap-around horizontal grille bars, tuning eye, 6 push-button tuning surrounded with plastic scutcheon, 4 knobs, 7 tubes, equipped for television conversion, standard broadcast, shortwave, police/amateur, AC. American, c. 1939

GRAYBAR RADIO, Model R7. Signed "Graybar Electric Co."
Tombstone design superheterodyne radio with matching polished veneer cabinet, accented with steel screws and embossed black trim; lower front window dial with escutcheon, upper three section cloth grille in matching gold, 3 carved wood knobs, BC, 8 tubes. American, c. 1932

EMERSON, Model 23-P, Limited Edition. Unique wood cabinet with base trim, two plastic tuning knobs, six panel matching cloth grille surrounds Disabled American Veterans' red, white and blue logo. "Lest we forget" is printed at the top of the logo; letters "D.O.V." are cut into the bottom of the escutcheon. The Model 23 was one of the most popular tabletop radios of it's time; the chassis was often used for limited edition promotional events. This special edition was produced to honor American Veterans, particularly from the First World War. BC, AC, 4 tubes. American, c. 1935

ZENITH TABLE RADIO, Model 7-S-529. Signed "Zenith Radio Corp., Chicago, Illinois." Chassis 7A02. Wood table model with front black dial, left wrap-around cloth grille with horizontal bars, 5 push-buttons, 2 knobs, 3 left and 3 right audio control toggles, 7 tubes, AM standard broadcast, shortwave, police/amateur, AC. American, c. 1941

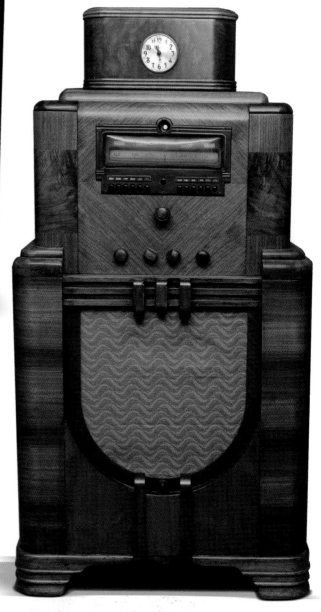

DE FOREST-CROSLEY CONSOLE RADIO MODEL SD 992.
Signed "Manufactured by Rogers-Majestic Corporation limited
Toronto Canada for De Forest-Crosley limited." Console with
sloping front, featuring 10 push-button automatic tuning,
8 tubes, 5 knobs, 15 inch speaker. American, c. 1940s

MAJESTIC MODEL 11656 FLOOR RADIO.
Signed "Majestic Radio and Television Corp., St. Charles,
Illinois." Console with automated tuning eye, 16 tube, 5 station
selector with programming timer located on top to enable the
listener to assign 24 hour pre-selected programming. Jensen
X-series 15 inch speaker. Completely electric, featuring
standard broadcast, shortwave, police/amateur, AC. Very
rare. Originally owned by a Majestic Radio executive.
American, c. 1937

MIDWEST RADIO CORPORATION MODEL 16-35, Signed "Midwest Radio Corp., Cincinnati Ohio" Console, wood, featuring dramatic chrome Deco-style escutcheon holding center frequency dial. 2 wooden knobs, 3 chrome toggles, 5 bands with 180 degree dial, back-lit dial indicates kilocycles and metric wavelength. Lower cloth grille covers 14 inch Magnavox 2000 ohm electrodynamic speaker, 18 tubes, AC. American, c. 1935.

MIDWEST '6-BAND' 18 TUBE, SHORT-BROADCAST AND LONG-WAVE RECEIVER. Signed "Midwest Radio Corporation, Cincinnati, Ohio." Console featuring chrome-plated levers, with advertised "Line-O-Light" feature that points out the frequency the set is tuned to. Six different tuning bands, covering from 125 kilocycles to 67 megacycles. The bands are lettered E-European Broadcast Band, A-American Broadcast, L-Low Frequency short-wave band, M-Medium, H-High, U-Ultra High, triple calibrated dial with 15 inch speaker. Some suspect the Midwest Company was guilty of "padding" their tube count in an effort to compete with the high performance, custom-crafted Scott radios which featured a large number of vacuum tubes. American, c. 1936.

STROMBERG-CARLSON MODEL 240-R RADIO RECEIVER, Signed "Stromberg – Carlson Telephone Manufacturing Company, Rochester, NY, U.S.A." Wooden console featuring half-round front case with curved double doors. Inner dial surrounded with metal escutcheon, cat's eye tuning indicator featuring 3 band "ranges" including broadcast, foreign broadcast and shortwave; 5 plastic knobs controlling volume, 2 stage-tuning, dial tuning, and power. Lower front horizontal wrap-around louvers. 10 tube chassis with 1 extra tube for "Magic Eye" (11 tubes total), AC.   American, c. 1937

ZENITH MODEL 12 S 475 FLOOR MODEL RADIO, Chassis 1207. Signed "Zenith Radio Corp., Chicago, Ill." Console, 12 tube, top-of-the-line receiver. Trademark black, shutter "robot" dial, automatic, standard broadcast, shortwave 1 & 2, featuring a waterfall design front cabinet with 7 vertical louvers and cloth grille over 15 inch speaker. Right station buttons, left tuning voice, BC. American, c. 1940.

GRUNOW TELEDIAL MODEL 1291 RECEIVER. Signed "Grunow Radio Company." Large art deco console radio with famous "Teledial." This unusual feature allows the user to pick the desired station using a rotary dial similar to a telephone. Known as the "Shirley Temple" radio because of the 1937 ad campaign for the radio, featuring the child star. Twelve tube chassis. American, c. 1937.

ZENITH MODEL 617 SUPER-HETRODYNE, Signed "Zenith Radio Corporation, Chicago, Ill. U.S.A." Ornate console superheterodyne radio with upper center escutcheon surrounding backlit dial. Featuring individual viewing windows for volume control (left), kilocycles (center), tone, bass and treble (right window). Dial light located above all three. Includes 4 wooden knobs for volume, frequency and base tone. Lower cabinet features dual 10 inch electrodynamic speakers covered in elaborately scrolled wood grille with matching cloth. 12 tubes, 6 carved standoffs. American, c. 1924

COLUMBIA CONSOLE PHONOGRAPH with KOLSTER RADIO.
Signed "Columbia Phonograph Company, Inc., New York, USA."
Lowboy walnut cabinet with top-load Columbia 78 rpm turntable
featuring custom brass tone-arm inscribed "The Audakco, New York."
Featuring custom designed Kolster radio situated above 15 inch loudspeaker covered with elaborate scroll cut-out and leopard design
mesh-grille cloth. Brass escutcheon surrounds 3 tuning knobs and
dial, 12 tubes. Left shelves hold permanent red and green colored
volumes, all titled "Columbia." Solid wood cabinet doors are carved
to look like hanging stage drapery. One of the first record players
powered by alternating current and certainly one of the most beautiful and distinctive high-end radios ever produced.   Inside top reads
"Viva~tonal – Columbia – like life itself." American, c. 1929.

RCA 'LOWBOY' RADIOLA,
Model AP-777-C. Signed "Victor Talking Machine Company, Radio Corporation of America, Camden, NJ USA." Elaborately carved lowboy console featuring central radio and early electric AC phonograph; 4 knobs in three clusters surrounded with extra heavy brass escutcheons. Inner phonograph, radio and controls situated behind decorative, intricately carved, inlaid double doors with brass latches. On/off toggle switch on left side of cabinet; top cabinet circled with ribbon of inlaid multicolored wood. Cabinet also features side storage for records. Four carved legs with additional horizontal scrolled pegs reinforcing legs; 11 tubes. Early electric record player using alternating current. Brass plate affixed to the back of the chassis reads "Warning, this instrument will operate only on alternating current." American, 1929.

SPARTON Model 301, THE NEW EQUASONNE
AC RECEIVER. Signed "The Sparks-Withington Co., Jackson,
Michigan, U.S.A." High-boy featuring double-front doors,
carved Italian design, inner dial, three knobs, 10 tubes, lower
back panel, AC. Original advertising caption reads "Radio's
richest voice." An excellent example of outstanding, elegant
cabinetry for the home. In 1928 Sparton claimed to be
fourth in industry sales (behind Crosley, RCA and top seller,
Atwater Kent.) American, c. 1929.

SCOTT POINTER-DIAL PHILHARMONIC CONSOLE RADIO MODEL 20A. Signed "Custom build by E.H. Scott Radio Laboratories, Inc., Chicago, U.S.A." Pedestal style, backlit dial, with two, tube powered, brass tuning eye indicators. Capable of reception from the standard broadcast band up to 80 megahertz, exposed 24 chrome tubes in the main chassis and another 6 in the power supply (30 tubes total), rare pointer-dial (instead of the beam-of-light which was used in later versions.) The radio frequency section of the receiver is sensitive and selective, and can easily be configured from the front panel to match reception conditions, built-in volume expander circuit and available with remote control. Other features include continuously variable tone controls, and a 60-watt audio output amplifier, all fed to a 15-inch loudspeaker. Custom built cabinet. Interior label reads "The finer things are always hand made." Considered the best radio circuitry of the day. American, c. 1937

GALLERY SIX
The Jones Gallery

The Jones Gallery

The vacuum tube represents the first great leap in the control and manipulation of electrons – essential to the development of electronics and radio technology. It was the vacuum tube that first enabled the human voice to be carried from San Francisco to New York (1914) – a feat that would have been decidedly impractical without the amplification the vacuum tube provided. Vacuum tube technology facilitated the first great advances in wireless telecommunications, as well as long distance telephony, the development of X-rays, the first computers, and the discovery and manipulation of radio waves, including radar.

The 20,000 assorted vacuum tubes in the Jones Gallery present the complete history and evolution of this wondrous technology, beginning with the Fleming Valve – considered the first vacuum tube ever produced (1904). Every era, every type, and every category is represented. Every significant manufacturer and country is included. The collection also features a variety of one-of-a-kind prototypes and specialty tubes (including the prototype of the image storage tube used on Lunar Orbiter III to take the first pictures of the dark side of the moon.)

The list seems endless: Early X-ray and rectifier tubes, De Forest transmitting and receiving tubes, laboratory prototypes for radar systems, even examples that trace the evolution of electron tubes from glass to metal. Every shape and every size (vacuum tubes as small as a child's finger and others as large as a kayak and weighing more than a hundred pounds); it's all here.

The following pages present only a sampling.

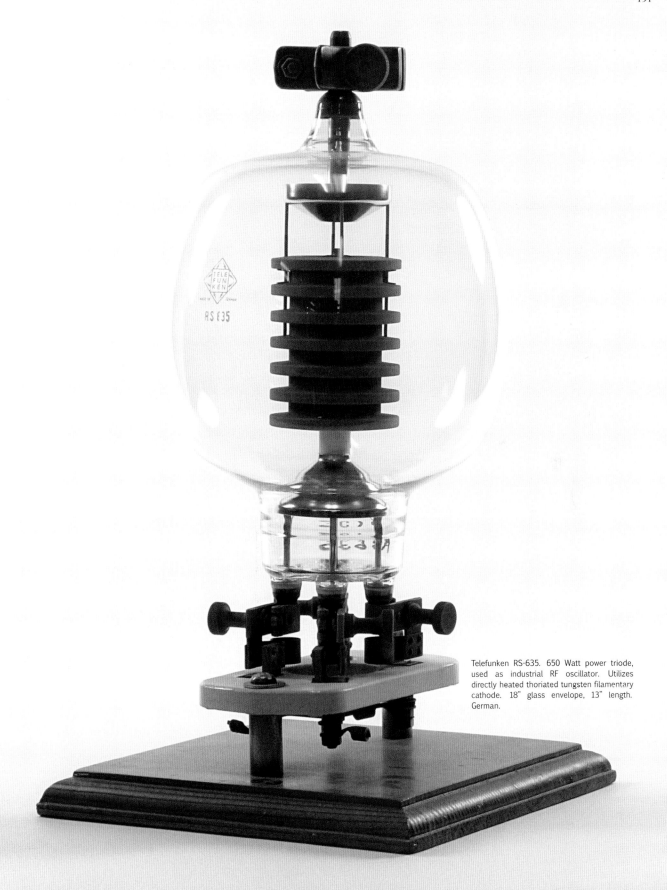

Telefunken RS-635. 650 Watt power triode, used as industrial RF oscillator. Utilizes directly heated thoriated tungsten filamentary cathode. 18" glass envelope, 13" length. German.

"...Vacuum tubes as small as a child's finger and others as large as a kayak and weighing more than a hundred pounds."

General Electric 862A RF Amplifier. Water cooled with 2 filament leads. Used in AM radio stations. 50 kW, long copper base would sit in a circulating bath of 15 - 25 gallons per-minute of water. Overall length is 63", glass envelope is 24", Copper water jacket is 39". 35 lbs. American, c. 1945

Tung Sol Lamp Works Inc. 5672 Sub-Miniature RF amplifier. Filament voltage of 1.5v, 65 Mw. Starting in 1907 Tung-Sol developed and produced the first commercially successful headlamp for automobiles. Approximately 1" in length. American, c. 1962

Tungsram 3V 25T-1 66-849-06 V110V.
Tungsram Radio Works, LTD, Westroad, Tottenham, London, England. 3V 25T and 3V12
T Rectifiers. Used as a rectifier for radio stations. Runs very hot and is both air cooled
(lower copper base) and water cooled (upper glass container). 22" length, glass envelope
20 dia., water cooling copper jacket is 10" in length, 38 lbs. English, c. 1963.

Heintz & Kaufman Gammatron Type 255X. Filament voltage 1.4v, 17" overall length, 19" dia. American.

General Electric GL316A - VT191 Valve Tube. For "ultra-high frequencies up to 750 MHz. Featuring close electrode spacings, reduced time of electronic transit, short heavy leads, low inter-electrode capacities. Primarily used for military. Advertisement reads "Go up as high as you like. The 316M takes you all the way!" 7" circumference. American, c. 1938

Gammatron Heintz & Kaufman Type 155. Heintz and Kaufman designed and built their own tubes, ignoring and later successfully challenging license and patent restrictions established by RCA. The tubes they produced were gridless Gammatrons. 14" length, 13" dia. American.

Federal ITT, "99". Components Division. 15" length, 20.5 lbs.

Machlett 846 JAN-CAGD. Suspended on springs in protective packaging. Water cooled oscillator and RF power amplifier. Copper anode, 5" glass dia., 9" length. American, c. 1935.

Ballast Tube. Regulating tube used to help keep the current constant drawing energy to keep the more sensitive tubes in the mechanism from burning too quickly; elongated wire. 5" length, 5" circumference.

RCA Radiotron 5671 Electron Tube. High powered, air cooled transmitting tube used in AM radio stations. Tubes like this were used in the 50,000 watt clear channel stations in the 1950s and 1960s. The filament takes an impressive 11 volts and 285 amps. Believed to come from KRVN Lexington Nebraska. The porcelain base is a chimney in which air from below is funneled up to the tube. 28" length of actual tube (without base), 34" overall length with porcelain base, 43" dia., 130 lbs. American, c. 1950s.

BTH B4
British Thomson-Houston Co. LTD
Early triode with tipped glass envelope highlighting smooth silver deposits. The B4 valve was intended primarily for low frequency power amplification. Steel base with stamped "A" and "G," 6.25" cir., 4" overall length. British, c. 1924.

CENCO 71266 Electron Beam Tube
Three element tube with an octal tube base. The glass envelope and inner disc are coated with a material that florescence when struck by electrons. Special purpose tube used to measure the ratio of the charge of an electron to its mass. 6.5" cir. at top, 5" overall length. American, c. 1930s

STC SY4019A Amplifier
Standard Telephone & Cables Co. Ltd
Navy-type triode audio amplifying tube. First established as a Western Electric Company (Australasia) Ltd, and later became STC in the mid 1920s. "SY" signifies Sydney and was used as a prefix to their type numbers. 7" cir. at top, 4" overall length. Australian, c. 1920s

Kellogg AC 401 triode amplifier
Kellogg Switchboard and Supply Company. An early triode developed for AC filament operation. Cylindrical plate in tapered shaped glass envelope; filament leads attach to the top (cap) of the tube (instead of the bottom). Bakelite base hot-branded with "K" logo and number. 5.25" overall length. American, c. 1926.

Marconi R Type Receiving Valve
Manufactured at Osram Lamp Works.
Well made early triode tube with tipped glass envelope featuring cylindrical plate mounted horizontally; stainless steel base. Sometimes called a "French R valve" though manufactured in England. "BBC" printed on glass casing.
6.25" cir., 4" overall length. English, c. early 1920s

Siemens Glz 40/6L
The filament-cathode connects to the thumb screws at the base. Crowfoot mounting provides the filament leads. Unusually heavy getter deposit on top inside of glass envelope. 10" cir. at top, 7.25" overall length. German, c. late 1920s

TRIGATRON SPARK-GAP VALVE 24B9. Signed "English Electric Valve Co. LTD ". Known as a"Sock" tube. Spark gap valve in a pressurized glass envelope, covered with web like sock held in place with heavy varnish to prevent explosion. First used in high altitude aircraft radar systems. The Trigatron helped establish and demonstrate the feasibility of navigation and bombing with radar. 115,000 Trigatron tubes were produced for the armed services between 1942-46. 6" length. British, circa 1942

Forced-air Transmitting Tube, type 6165. This very rare, co-axial forced air UHF transmitting tube is gold plated with no identifiable manufacturer. 5" dia., 6.5" length. American.

Micromesh R.2. Amplifier
Standard Telephone & Cables Co. Ltd
Triode with unusual tilted grid and plate assembly. Original blue and white paper sticker on the glass envelope, 1" square with scrolled "Micromesh" in center.Inner glass is opaque silver from getter residue. 2" dia., 4.75" length. English, c. 1930s

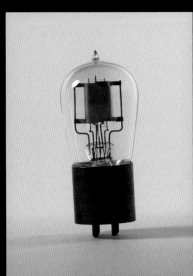

Cunningham, Inc. 201 Amplifier/Detector
Early 201 triode used as an amplifier/detector or amplifier. Globe-type glass envelope suggests a "soft" vacuum (no getter) design. "Licensed for amateur, experimental use only." American, early 1920s

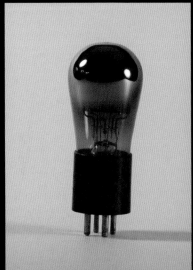

Super Airline Gx-201M Amplifier
Triode with marbled Bakelite base. Gold colored deposit on the inside of glass envelope results from the getter material (phosphorus & magnesium) used to absorb any gas that remained in the tube after evacuation. American, c. mid-1920s

RCA Radiotron UV-201A
Triode featuring interior glass envelope coated with silver deposit created by the combination of the getter materials within the tube during the manufacturing process. prongs were a UV designed by RCA. Used for amplification. American, c. 1921

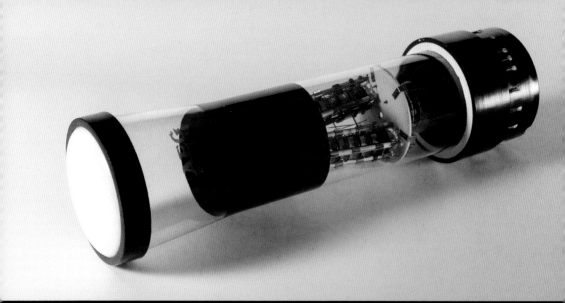

AEG HR2/100/1.5A CRT, Allgemeine Elektricitäts-Gesellschaft, General Electricity Company
Wehmacht markings, with 2 complete electron gun/deflector plate assemblies and 18 contacts around the base. Measures 4" across tube face, 15.5" overall length. German, c. 1960s

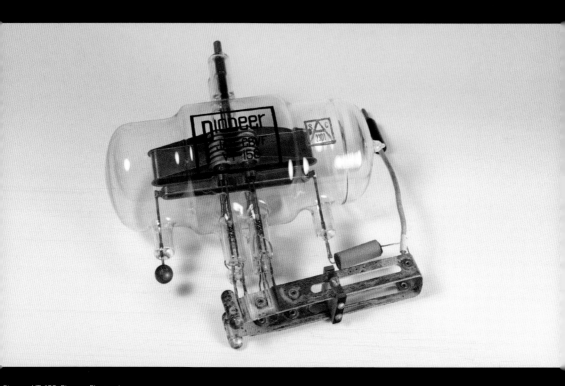

Pioneer VT-158, Pioneer Electronics.
Four-triode tube manufactured by Pioneer just after WWII, and first developed by EIMAC. The two thick plate studs at the top form a 50-ohm balanced line to extract the maximum power output. EIMAC, Heintz & Kaufman, and Pioneer Electronics were the most significant manufacturers of this fascinating and rare radar tube, also known as the Zahl tube, named after its inventor, Major Harold Zahl. 7.5" wide, 10.5" dia., 9" overall length. American, c. 1945.

The "Moon Tube"

TELEVISION LINE SCAN TUBE PROTOTYPE. Signed "CBS." The Line Scan tube was developed by CBS Laboratories, and manufactured by the Jet Propulsion laboratory, for film scanning application. The electron beam is deflected in only one direction, so that it produces a single line on the phosphor within the tube. Used with Lunar Orbiter III to photograph the dark side of the moon. The system took 40 minutes to read out a single exposure which consisted of 1700 horizontal scanned lines. 20" length, 11" across top, 13" circumference, 3 lbs. American, c. 1968

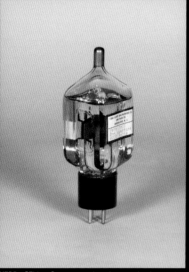

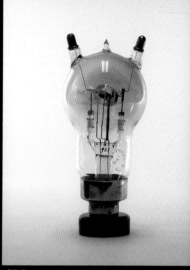

KERR CELL. Square and substantial clear glass envelope holds yellowish transparent nitrobenzene fluid; with metal cap and four prong base. Named after John Kerr, Scottish physicist and ordained minister. The Kerr electro-optic effect allows electricity to control light at an extremely high speed. Kerr Cells were utilized in early television instruments as a means of modulating a beam of light electrically. White and blue paper label reads: "TUBE LIGHT ENGINEERING CO. 300 PARK ST. MOONACHIE, N.J." Typewritten text reads, "Kerr Cell 2mm Aperture 4000 volts" Measures 6 x 2". Extremely rare. American, mid 20th century.

Thomson-CSF TH5T1000A - BW38. 1000 Watt transmitting tube, 9" length, 15" dia. British.

SIF Transmission triode with 75 Watt transmitting power. Considered too expensive for amateurs. 7" length, 10" dia. c, 1935.

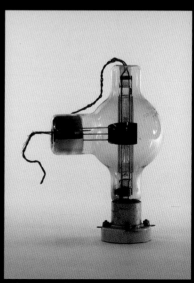

Vacuum Capacitor. Vacuum capacitor uses high vacuum as the dielectric (instead of air or other insulating materials) allowing for a higher voltage rating and greatly reduces the chance of arcing between the plates. These tubes were used mostly in high powered broadcast transmitters, amateur

RCA Radiotron UX-867 Photo Tube.
The Radiotron is a argon gas vacuum tube and made possible the development of radio and the broadcasting industry, including television. 6" overall height, 7" dia. bulb. American, c. 1920s

U.S. Navy Type 860 Transmitter Tube.
Early transmitter tube also known as the 860, with oversized elongated grid. 8.5" height.
American, 1929."

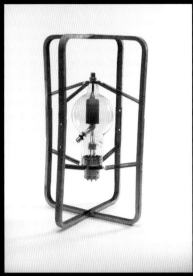

EIMAC VAC Tube 450TL. 450 watt triode. Shown in its shipping spring mount to avoid damage. 12" overall height, bulb circumference 16". American, c. 1943.

High Voltage Solenoid. 23" overall height, 12" dia. Corrugated glass cylinder. Ceramic insulator 8", glass 14.5". 23 lbs.

Unidentified, but likely a high-voltage power rectifier. Probably intended for forced-air cooling, although some such tubes were often immersed in cooling / insulating oil. Cooling fins, cast aluminum, base, plastic with metal sleeve. 16" overall length, 15" glass. 12.5 lbs.

Western Electric 101D Amplifier "Tennis Ball" Tube. DC Amplifier Triode, used in early telephone switching stations and very early audio amplifiers. American, c. 1921

De Forest long plate Spherical Audion. Considered the first American electronic component and one of the first of its type anywhere. In 1906, Lee De Forest created the "Audion" vacuum tube by adding a grid element to the Fleming Valve creating the first triode vacuum tube. The Audion could amplify a signal, making way for the invention of many advanced electronic technologies including long distance telephone, radio and television. American, c. 1914

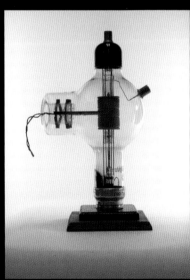

Westinghouse VT-19 Tetrode, JAN-CWL-861 17" overall length, 22" dia. American, c. 1929.

Museum curator Jonathan Winter with Federal Telephone and
Radio F-660 RF Amplifier. 30 kilowatt Rectifier tube with triode.
23" overall length, glass envelope is 11"; copper water jacket
and plate is 10", 17 inch dia. glass. 13 lbs. American.

UV212 Magnetron
This very rare tube is the first commercial magnetron, used to produce microwaves.   5" dia., 6.5" length.
American, c. 1921

Western Electric Co. Cathode Ray Tube. Early CRT with electrostatic deflection. American, c. 1924.

RCA Radiotron 1850A Electron Camera Tube/Image Iconoscope. One of the earliest commercially available camera tubes. The earlier version, (RCA 1850) dates back to 1939. The selling price in 1948 was $540 (In 1948 you could buy a house for $3000.) Electron Gun (neck) 8" long, dia. of glass envelope is 22".
American, c. 1950

AL JONES

There are many ways to judge a collection of vacuum tubes: range of technology, age and date, national origin, size, scope, historic significance, et cetera. By any measure the collection developed by Al Jones is one of the most comprehensive ever amassed.

Like most tube collectors, Al began as a radio amateur (W1ITX) when vacuum tubes were an essential part of communication. Al started collecting older ham radio tubes and "Got bitten by the collecting bug." He was soon gathering transmitting, receiving, medical and specialty tubes from all over the world. He is the founding president of the Tube Collectors Association, and in 2004 received their Schrader Award for outstanding achievement in developing a comprehensive collection. Al has received the Antique Wireless Association's Tyne Award in recognition for his exceptional contribution to vacuum tube history.

In 2003, Al graciously donated his magnificent collection of 10,000 perfectly preserved and catalogued vacuum tubes to the Museum. Combined with the Museum's already significant collection, they create one of the most complete and important assemblages of vacuum tube technology ever displayed and preserved.

Al Jones in his office, holding his "Moon Tube."

GENERAL INDEX

(SEE APPARATUS INDEX FOR REFERENCES TO APPARATUS PICTURED)

APPARATUS INDEX

INDEX TO APPARATUS PICTURED

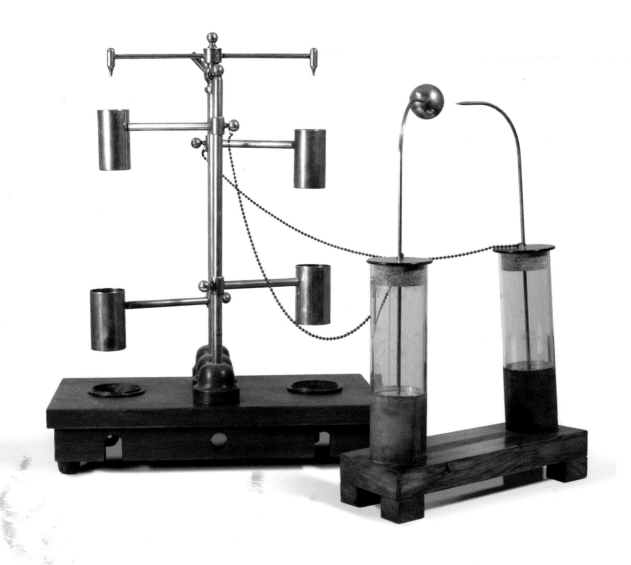

KELVIN'S THUNDERSTORM. Signed "Angelo Arrighini." In 1867, Lord Kelvin (William Thomson, 1824-1907) demonstrated that static electricity could be produced from dripping water. His experiment consisted of a metal tank filled with distilled water suspended above a table.   The bottom of the tank contained two drains with taps to control the flow of the dripping water. Below each tap was a metal can that collected the water. A copper loop was placed just below each tap, between the tap and the can, so that water would drip from the tap through the loop and into the can below. The loops and cans were cross-connected with a wire (i.e. the loop under the tap on the right was connected to the can under the tap on the left, and visa versa.)

As the water drops through the air towards the collection can, it obtains a small static charge through induction. Over a period of time the charges build up in the Leyden jars until a spark occurs at the gap. Italian, c. 4th Qtr 19th Century.